AF597406

THE ROLLING STONE® GUIDE TO HIGH FIDELITY SOUND

by Len Feldman

Library of Congress Catalog Card Number: 73-90733
ISBN: 0-87932-070-2

First printing

Straight Arrow Books
625 Third Street
San Francisco, CA 94107

Distributed by Quick Fox Inc.
33 West 60th Street
New York, NY 10023

Order number: 102070

Production by Planned Production

Printed in the United States of America

To Rayma

Contents

The actual model used in the original RCA trademark. (Courtesy of John T. Mullen, Curator, AES Museum Exhibit, "80 Years of Recorded Sound")

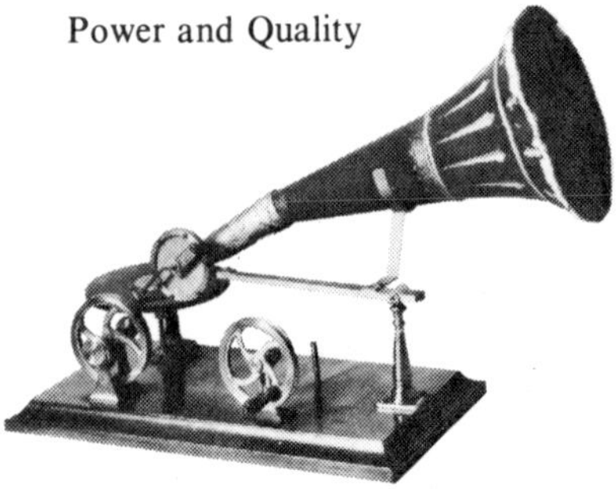

Preface

Ever since Tom Edison bellowed "Mary had a little lamb" into his first hastily conceived and produced wax cylinder record-making machine back in 1877, people have been writing books explaining and extolling the virtues of good sound reproduction in the home. Time and technology have improved reproduced sound quality over the years, but most of the books on the subject have followed a rather dreary pattern of trying to make audio engineers (or at least audio technicians) out of their readers. Like many other hobby-oriented pursuits, the quest for good sound in the living room has become buried in the mire of technical jargon, hard-to-understand specifications, myths and even hi-fi folklore which tends to frighten and discourage the audio neophyte who simply wants to put together a good sound system for his or her personal enjoyment. I've tried to avoid that approach in this book. Certainly there are performance characteristics of the different components of a hi-fi system that can help you choose a "good" system, and these are explained and given as much emphasis as I feel they require. On the other hand, there are many terms and specifications that are of primary interest to the audio designer and manufacturer and are of little or no use to the prospective buyer of hi-fi equipment. These I've ignored, preferring rather to concentrate on *how* you go about putting together a fine hi-fi system, rather than on *why* one manufacturer chose to use one kind of design approach as opposed to another.

I've also tried to separate the myths from the reality—the ad agency copy from the facts. Where I've used photos and illustrations of actual equipment, the intent is never to plug a specific product, but rather to illustrate a point or feature that's common to a great number of good products. The fact is, there *are* a lot of good component hi-fi products on the market today—but there's also a lot of junk around that parades as "high fidelity" equipment. If you know where to go, you can buy little plastic emblems that say "High Fidelity" for under a penny apiece. They all come with self-adhesive backing and can easily be pasted on to the front panels of any piece of electronic garbage (or lipstick containers, or broiler ovens or TV sets—which by the way offer about the *lowest* audio fidelity around). The purpose of this book, then, is to help you pull off the labels from the phony equipment and stick them on components that merit the much-abused descriptive term—high fidelity.

Len Feldman
March, 1974

Chapter 1

INTRODUCTION–HOW SOUNDS REACH YOUR EAR

Sound is a term used in two senses: it signifies the sensation experienced by our ears and, in a more objective sense, it refers to the vibration or transmission of vibratory energy that creates that subjective sensation. It's this dual meaning which probably first gave rise to the age-old argument about whether or not "sound" is produced by a falling tree in an unpopulated forest. We'll leave that one to the philosophers, since we are more concerned with the energy vibrations in the objective definition of sound.

If you pluck a guitar string, the air in the vicinity starts to vibrate in alternate compressions and expansions which are called sound waves. The analogy sometimes used—that of comparing sound waves with the waves set up in a pond when a pebble is dropped into it—is useful but only up to a point, since water is an incompressible fluid whereas air is a highly compressible gas. If you pluck the guitar string harder, the compressions and expansions are greater and the sound is louder. If you permit only a part of the string to vibrate, the number of vibrations per second will increase and this is termed an increase in pitch. Actually, the string of a guitar, when plucked, sets up vibrations of many related frequencies (or pitch) all at once. There is the *fundamental* frequency of a given intensity and there are secondary frequencies—all multiples of the fundamental and all of lesser intensity, which combine to make the instrument sound like a guitar (or a piano, or a violin or any other instrument). This combining of harmonically related tones is illustrated in Fig. 1-1 and is what determines the quality or kind of sound we hear.

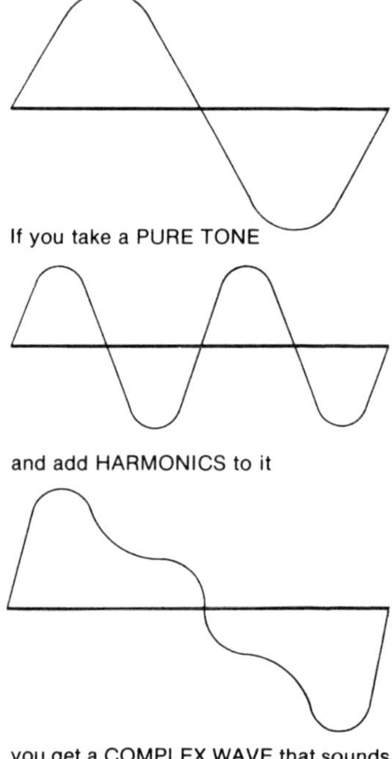

Fig. 1-1 Musical sounds consist of fundamental pure tones modified by harmonics.

Different combinations of fundamental and harmonically related tones are what make various musical instruments sound different from each other—even when all of them are playing the same basic "note." It's generally accepted that people hear tones from about 20 vibrations per second to roughly 20,000 vibrations per second. Since vibrations-per-second is quite a mouthful to repeat over and over again, audio-involved people first used the term cycles-per-second (abbreviated CPS) or kilocycles (thousands of cycles per second, abbreviated KC), but more recently, the abbreviations Hz and kHz have become universally accepted (in honor of a mathematician named Hertz, who first expounded these and other "wave motion" theories).

Sound waves travel at about 1090 feet per second in air (much faster in water and faster still in a solid medium), but they get weaker and weaker as the distance from the sound source increases. In fact, they get weaker as the *square* of the distance from the sound source, so that the sound heard forty feet away from our singer in Fig. 1-2 is only one-quarter as intense as it would be if the listener were standing twenty feet away. This relationship holds true outdoors, where the sound waves are free to travel in all directions.

In a room, however, sound waves bounce off the walls, ceiling and floor and some of the reflected sound energy returns to the listener from all these surfaces. In a small room the reflections occur so fast that you

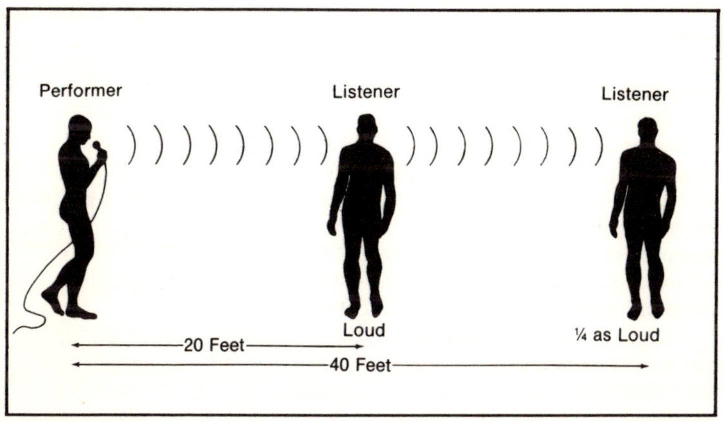

Fig. 1-2 Sound intensity varies inversely to the square of the distance from the sound source.

can't mentally separate them from the original sound. In a larger room like a concert hall there's enough of a time delay between the original sound and the reflections to give a "reverberance" effect—the sonic effect that *tells* your brain that you are in a large auditorium or hall. If the room is made still larger, the effect becomes an "echo" and you actually hear the sound repeated one or more times until it dies away.

The act of "hearing" is in itself a very complex process, but we need take it only as far as the eardrums in both your ears, which are caused to vibrate when those air compressions and expansions reach them. If you get far enough away from the source of sound you want to hear (especially outdoors), you reach a point where you can no longer hear that sound. This brings us to the upper portion of Fig. 1-3 in which the principle of sound *reinforcement* is shown.

When the microphone is placed close to the singer it picks up the complex vibrations produced by his vocal chords (which set the air into vibration) and transforms those vibrations into minute electrical signals that are an exact replica of the *physical* vibrations impinging on the microphone—assuming that the microphone is able to accurately make that transformation. Once in this electrical form, the signals can be amplified by an amplifier, fed to a loudspeaker and transformed back into mechanical vibrations to set the air in motion once more. If all parts of the system introduce no alteration of the *shape* or tonal content of the singer's original efforts (and that is what high fidelity is really all about), then the only change that takes place is one of *intensity*—the sound produced by the loudspeaker is much louder than that produced by the vocalist. As a result, it can be heard at greater distances from the vocalist than would have been possible without the aid of sound reinforcement equipment.

Live Concerts Aren't Live

Aside from a few exceptions such as opera performances, symphony concerts and small groups performing in small halls, just about any concert you attend—though billed as a live performance—is not live at all. You may be *watching* live people, but you're *hearing* the output of sound reinforcement systems, specifically, the loudspeakers of those systems. Modern rock groups often use sound reinforcement systems not only to amplify the sound of their voices and instruments but also to alter the tonal quality of the sounds they make. This is a perfectly legitimate use of electronic sound equipment in that the equipment becomes part of the performer's instrument kit; it is *not,* however, an end goal of home high fidelity components. Their job is

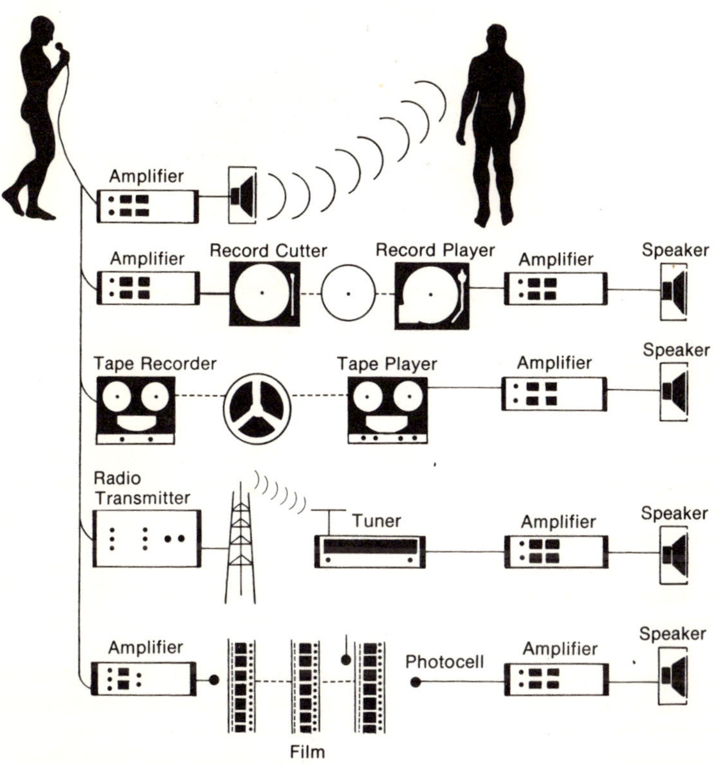

Fig. 1-3 The many ways in which sound can get from *there* to *here*.

Microphones—the beginning of the recording process. (Courtesy of John T. Mullen)

to attempt to reproduce the sound of a concert with the same tonal quality you would have heard if you had attended that concert—whether the concert consisted of "live" *or* electronically enhanced sounds. In brief the components should not contribute any sonic content that you wouldn't have heard at the original performance.

Getting back to our diagrams of Fig. 1-3: if you move the listener further and further away from the performer (like miles or continents away), no amount of sound reinforcing equipment can make that performer loud enough to be heard by the remote listener. At that point, you have two basic choices: you can either *store* the music on one of several available mediums for subsequent reproduction in your home at your convenience, or you can have the performer broadcast his performance by radio, in which case you can hear a reproduced version of the performance shortly after it takes place. (Even radio waves take some time to travel from one place to another—although at 186,000 miles per second, it's rather academic.)

As you can see in our diagram, there are basically three ways in which to store the electrical equivalent of sound programs. The electrical microphone signal can be amplified and used

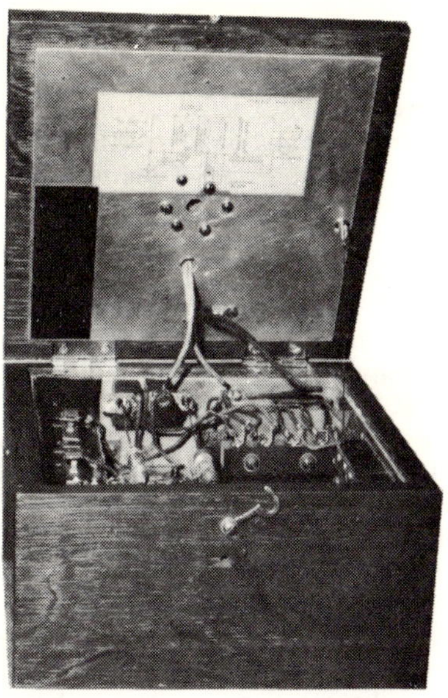

The microphone preamplifier was too bulky to be portable. (Courtesy of John T. Mullen)

to process a recording disc, in which the wiggles of the groove correspond to the original sound waves; the signal can be used to magnetize a tape running through a tape recorder, in which the signal is translated to magnetic patterns on the tape that correspond to the original program; or the electrical signal from the microphone can be amplified and cause a lamp to flicker, changing the exposure pattern on a strip of film (the way generally used to get the sound into moving pictures).

For "almost instantaneous" radio reproduction the amplified microphone signals can be used to modulate or vary the radio signals sent out by a radio station transmitter, either AM or FM; we'll get into the differences later on.

On the listening end, you can play the record and hear the program that way, you can play the recorded tape or you can turn on your AM or FM radio system. I suppose you could run a movie film through a sound projector, too, but that's not an awfully popular way to play hi-fi sound at home and is best left to professional filmmakers. Although there are books available on filmmaking and sound-on-film this is not one of them. Essentially, we'll be dealing with the three centrally shown music reproduction systems in Fig. 1-3. If you want to be a recording engineer, a broadcast engineer or a disc maker, you'd better go out and buy another book.

Components Versus Package Consoles

For all the advertising and promotion of high fidelity components that's around, you may be surprised to learn that the "component" segment of the hi-fi business constitutes less than fifteen percent of the home-sound entertainment business. If you consider the fact that more dollars are spent on TV sets than on the whole category of home music systems, you realize that the component hi-fi market is small indeed. But it's growing—rapidly. After more than a quarter of a century of promotion of the virtues of components, the message is finally getting through. It's the young people who have been turned on to good sound and it is they who won't settle for anything *less* than component sound quality.

When you try to pump sound out of a loudspeaker system, as we've already seen, that loudspeaker has to set up vibrations, that is, it must move air around the listening room. If you were to mount loudspeakers in a piece of furniture in which you've also mounted the delicate record playing mechanism with its fragile tone arm and pick-up cartridge and stylus, the vibration created by the loudspeakers would be transmitted right up through the cabinet. Try to get loud sound levels from such an arrangement and those vibrations become strong enough to make that record tone arm jump right out of the groove or, at very least, cause a grumbling growl that engineers call *acoustic feedback*.

So how come the "French provincial" and "Spanish modern" one-piece wonders *don't* take off when you play them? Simple. The engineers who create these eyesores are just as aware of the problem as their component-designer counterparts and they take steps to prevent such sonic catastrophe. They simply limit the bass, or low-frequency response, where most of the sonic energy and vibration take place. By so doing they also cut out a most important portion of the music, but it *does* prevent rumbles, growls and "groove-skipping" when playing records. Usually they limit the loudness as well (all phony power output claims notwithstanding) so that you never get to hear the music at anywhere near the loudness level of the original performance.

The first requirement of a true component high fidelity system is that the loudspeakers be completely separated from the rest of the system (and particularly from the phonograph portion). The moment you free the speakers from the record player, designers regain the freedom to incorporate full frequency response. That means accurate reproduction of *all* audible frequencies, from 20 Hz to 20,000 Hz. Accuracy of musical reproduction is the best definition I know for what a component high fidelity system is supposed to accomplish, but that's really only one of the many advantages of a component system over the "orange crate" approach.

Non-Obsolescence

If you owned a console radio-phonograph back in the late 1950s when stereo records first made the scene, you were out of luck. You could go on playing your single-channel records and hope that their antique value would increase or you could throw out the entire console package and start all over again. If, on the other hand, you owned a component hi-fi system, you didn't have to discard a single piece of it. You needed an extra loudspeaker and amplifier for that second stereo channel and you probably had to buy a new phono cartridge which provided the necessary two signals from the record grooves; but you didn't discard anything you already owned in terms of the electronics and speakers that represented the major financial outlay of your system.

Now everybody's talking about four-channel sound and this new, more realistic hi-fi medium is beginning to take hold. Again, if you're stuck with a majestic mahogany monster, there's no way for you to convert it to quadraphonic sound. If you already own a component system you can easily update it and turn it into a state-of-the-art four-channel hi-fi system. You'll have to add some components (and we'll get into that in Chapter 11), but you don't have to throw away anything you already own. As a pure investment, then, components make a lot more sense.

Flexibility

Then there's the matter of flexibility. Suppose you decide that you'd like to have good sound in a second or even a third room. If you own a console package you might take the back off (if it has one), grab your soldering iron, a bunch of wire, some switches and other assorted parts. Then wire up those extra speakers and provide proper switching facilities. Alternatively, you can go out and buy a second and third console and put them in the bedroom (if there's enough room) or in the den. Then when the next major improvement comes through you'll have storage space for shirts and things.

If you own separate components, all you have to do is buy a second pair of speaker systems. Your electronics provides for these extra speakers and even has switching facilities so you can play one system, the other or both systems together. You can start with just record-playing facilities and add tape or FM radio later on. You can even feed the sound of your TV set through the better electronics of your components and discover that the bad sound of TV's comes from manufacturers in order to keep TV prices down. If you really want to get involved, you can add a whole family of relatively inexpensive accessories to your basic electronics. Such accessories give you better control of tonal response and artificial reverberation. You can even add four-channel adapter devices that help you combine your existing electronics with the extra electronics you'll need when you're ready for quadraphonic sound.

In short, the component approach to hi-fi has a building-block model that lets you build a system as simple or as complex as you want. It lets you change your mind as you go along with the *least* strain on your budget.

There's another thing you ought to know about hi-fi components, their makers and their method of distribution: most manufacturers of high quality components are relatively small companies. More often than not they were started by hobbyists who loved good music and developed equipment for their own personal satisfaction. Later they figured that if they enjoyed the results, others less technically skilled might like the same kind of good sound in their homes. As small companies their distribution pattern was—and is—a "one-step" affair—from manufacturer to dealer.

The giant electronic companies who continue to sell console-type equipment generally employ a "two-step" system of distribution, which means that there's a distributor "middle man" who has to make his profit on the item before it gets to the dealer from whom you buy. That extra mark-up amounts to between fifteen and twenty percent and obviously results in your getting less for your money in the long run even if you ignore the cost of the "furniture" that contributes absolutely nothing to the sound you end up hearing.

Discount Shopping

As you start making the rounds of hi-fi component shops in search of your dream system, you'll soon discover that prices for a given component are not the same in every shop. Some stores offer what seem like sensational discounts while others stick to the manufacturer's printed prices. Complete systems (in which the dealer has arbitrarily "put together" a system of components) often seem to have price tags that offer even greater savings.

Where you shop really depends upon what kind of service you want. If you feel you know what you want and are able to walk into a discount store and pick up your purchases in sealed cartons, there's nothing basically wrong with shopping for the best price in town. Product warranties are given by the manufacturer rather than by the dealer, so you don't even lose out there. On the other hand, dealers offering smaller (or no) discounts generally offer all forms of extra service, the chance to audition the equipment you plan to buy in properly furnished sound rooms and dealer-level warranties which insure faster repair or replacement service in case anything goes wrong during the warranty period. Dealers are all charged pretty much the same price by manufacturers (depending upon quantity ordered), so it stands to reason that if one dealer offers lower prices he's working on a smaller profit margin and must be cutting his overhead by reducing customer service one way or another.

Let's Keep the Concert Hall Where It Belongs

Before we get into the details of mono, stereo or quadraphonic hi-fi and how to put together a system, let's dispel one of the most "ancient" myths in the business. It used to be fashionable in hi-fi circles to say that the object of hi-fi was to bring the concert hall into your living room—to make you think you're attending a concert when, in fact, you're in the comfort of your own home. Today, most of us know better. For one thing, you can't play music in your home as loud as it is played in a concert hall (though some people never stop trying). For another, when you attend a concert, there's much more than just music involved. There's the feeling of *being* in a large hall, surrounded by an enthusiastic audience. There's applause and there's emotion and there are performers to look at and admire—psychological factors that no electronic system can ever duplicate. A good hi-fi system should bring music into your home and leave the concert hall downtown where it belongs.

Chapter 2

FROM MONO TO STEREO TO QUADRAPHONIC

It all started in the early 1930s when sound was added to motion pictures. Radio and phonographs were popular long before that in homes but no one paid much attention to the faithfulness or fidelity of musical reproduction attained by these early products. It was considered enough of a miracle just to have such music available at home. In the case of movie sound systems, cost was less of a problem and audio engineers engaged in this new work were able to design and build sound systems that were vastly superior in quality to the "squawk boxes" then in use in homes.

Many of these engineers and technicians, aware of what good sound could really be, put together systems of their own for use in their homes, and the high fidelity component industry was born. Friends heard this better sound and wanted it too, and many of the early hi-fi component companies were founded by just such technically oriented sound engineers who dedicated themselves to spreading the good sound.

After World War II things began to move pretty quickly. The quieter (less surface noise) longer long-play record was introduced by Columbia Records in 1948 and FM broadcasting and reception (capable of higher fidelity and less static and background noise) served as an added inducement for people to have good sound equipment in their homes. Now there was an ample source of good program material that justified the expense of high fidelity equipment. By the early 1950s, owning a hi-fi component system had become something of a status symbol. Enormous speaker enclosures occupied a goodly portion of living room floor space, but since only one of these boxes was required in that monophonic era people tolerated the unattractive speaker boxes in order to hear good sound. High fidelity components of the era were often ungainly looking things too, with glowing vacuum tubes exposed and no attempt made to integrate these devices into the living room environment. At the same time, the large mass-production companies were producing "radio-phonograph" consoles by the millions and poking fun at the "hi-fi nuts" who were crazy enough to get involved in cables, wires, soldering irons and the like. One of the ads run by one such "package producer" likened the selection of a set of hi-fi components to trying to buy a piano by choosing a Steinway keyboard, Baldwin strings and a Knabe piano bench. Of course the analogy is ridiculous and today that very same manufacturer is belatedly trying to enter the "component" market—some twenty years after belittling it. Even in the early days of component hi-fi systems interconnection of a tuner, record player, amplifier and speaker was quite simple; but with limited advertising funds available component makers had a hard time convincing music lovers of this truth.

Limitations of Monophonic Sound

Fig. 2-1 shows what happens when we attend a live concert. Sounds from each group of musicians reach the listener from a variety of directions. Because of wall and other reflective surfaces some of the sounds that originated from stage-left actually reach the listener from the right and vice-versa. If you place a single microphone in front of the stage, as shown, all the direct and reflected sounds will be picked up by that microphone. If the signal from the microphone is used to make a recording that is played back over a single-channel (monophonic) home music system, the effect on the listener is as shown in Fig. 2-2: just about all the sound in the recording will reach the listener from what amounts to a point-source or single point of origin—the front of the loudspeaker itself. It's as if you are listening to the music through a small hole in the wall. However good the equipment may be, the sound lacks the feeling of a live performance and can best be described as "flat" or "confined." Most listening rooms in homes don't do much to restore the ambience of a large hall because, acoustically speaking, compared to large auditoriums, living rooms are relatively "dead."

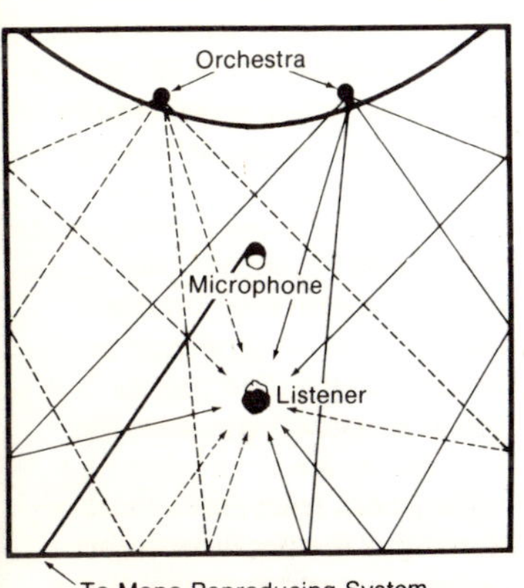

Fig. 2-1 When you attend a live concert, you hear direct and reflected sounds.

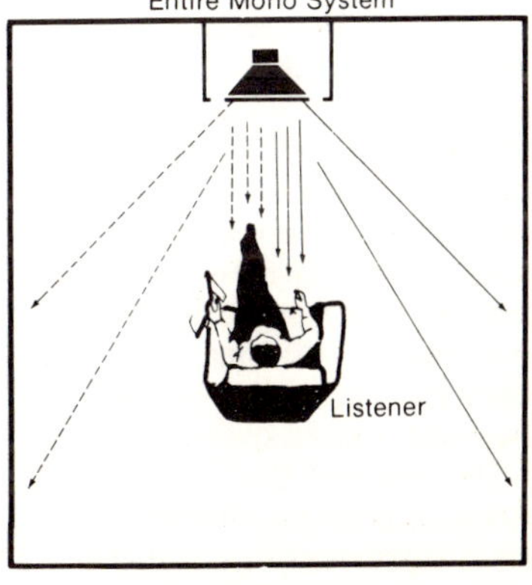

Fig. 2-2 In a mono system, all sound seems to come through a small hole in the wall.

Binaural Sound

In the early 1930s Bell Telephone Labs conducted a series of demonstrations involving binaural sound. It had long been known that our ability to localize sound stems from our being equipped with *two* ears spaced about six inches apart and facing in opposite directions. Even though sound travels at about 1000 feet per second (in air) and the time of arrival of a sound coming from your left differs by only about .00046 seconds from left ear to right ear, you are easily able to pinpoint the source of the sound as having come from the left, as illustrated in Fig. 2-3. The sound intensity reaching your right ear under these conditions is also diminished, since your head itself is in the way.

If two microphones were positioned on a "dummy" head to duplicate our ears and if recordings were made of the sounds picked up by each microphone separately and played back over separate earphones worn by the listener, every sense of space, direction and "liveness" could be captured and reproduced. Such experiments have been conducted many times over the years and the effects can be quite startling.

The results of these experiments suggest that all real-life sonic experiences can be duplicated with just two channels of sound. Indeed they can be—provided that the channels (microphones and earphones) are arranged for binaural sound, as described above.

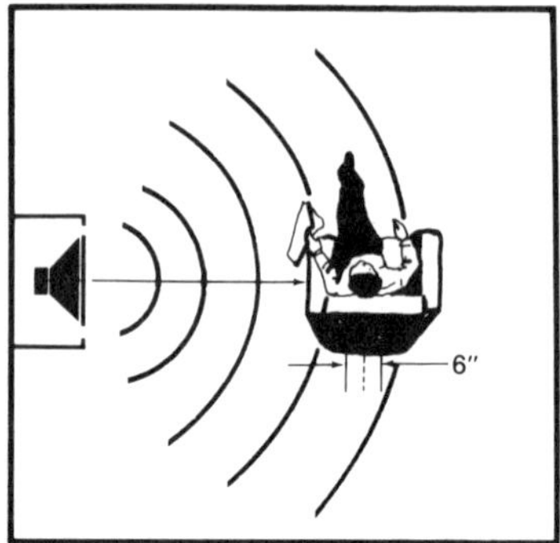

Fig. 2-3 You know the sound comes from your left because it reaches your left ear *first.*

Stereo Listening

In a typical stereo component listening set-up, however, speaker systems are usually spaced widely apart. In order to create the illusion of space the microphones at the recording end of the system have to be widely separated too. After all, in a stereo set-up, *each* ear of the listener is exposed to sounds coming from *both* left and right speaker systems. Thus much more exaggerated *separation* techniques have to be used at the recording end to providc a "plane of sound" sensation for the end listener.

The switch from mono to stereo listening really began with the introduction of the stereo disc in 1958. By inscribing "left" program information on the left wall of the record groove and right information on the "right" wall, it is possible to design a phono pickup and stylus which generates two separate electrical signals, each of which can be fed to a separate amplifier and a separate speaker. Furthermore, the "separation" between the two signals need not be total because even when a musical signal is just a bit louder from one speaker than it is from the other we tend to "hear" it as coming from the louder sound source. This idea will be important when we talk about certain kinds of four-channel or quadraphonic discs now gaining in popularity.

Obviously, recording a program stereophonically, as illustrated in Fig. 2-4, can increase the concert-hall feeling considerably. As shown in Fig. 2-5, the sounds of an orchestra or performing group are no longer beamed to the listener from a single point; instead sound is dispersed across a line stretching between the two speakers, creating a sort of curtain of sound. It's even possible to "place" a recorded sound mid-way between the speakers—an effect that occurs when identical and equally loud sounds come from both the left and right speaker systems.

The popularity of stereo hi-fi systems in the 1960s was not purely the result of the introduction of the stereo LP disc. In 1961 the Federal Communications Commission, after a study of several years, changed the broadcasting rules for FM and approved a system whereby the two different signals of a stereo program could be broadcast simultaneously over a single station. A new kind of radio tuner or receiver was required to unscramble the two signals. It was quickly made available to the

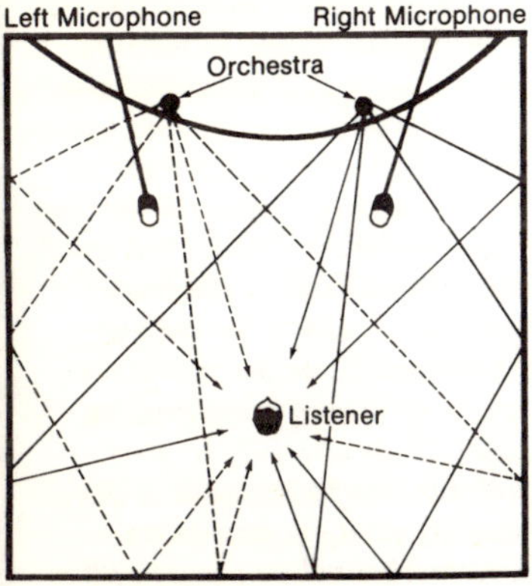

Fig. 2-4 Two microphones are used to record a live concert in stereo.

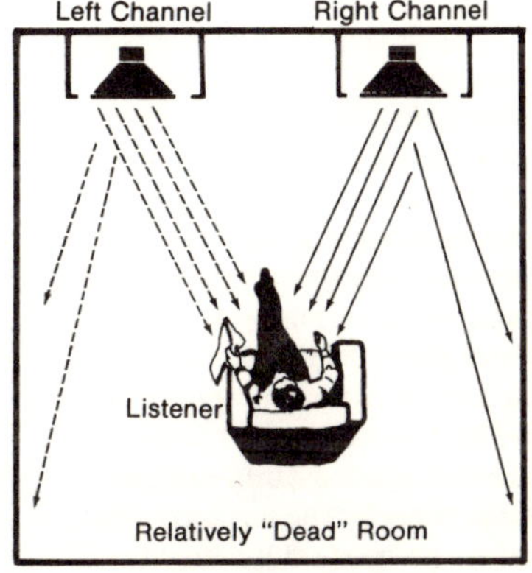

Fig. 2-5 Stereophonic listening provides a sense of left-right directionality.

public—first by the better component hi-fi producers; later by the mass-merchandising console makers. Today any tuner or receiver worth having will be equipped with circuits capable of receiving two-channel stereo broadcasts. Just about half of all the FM stations in the U.S. transmit all or most of their programs "in stereo."

Just as stereo was gaining in popularity another technological breakthrough occurred in audio products. The transistor, invented in 1949, finally was put to practical use in the circuits of hi-fi amplifiers, tuners and receivers. Because transistors are small in size and produce far less heat than vacuum tubes, today's amplifiers and receivers are no larger in size (and often smaller) than the single-channel vacuum tube amplifiers and receivers that were popular in the early days of hi-fi.

A giant step beyond the transistor is the *integrated circuit,* often abbreviated as *IC.* These little "chips," often no larger than a fingernail, contain dozens or even hundreds of transistors, further reducing the number of parts required in an amplifier or tuner. We're really at the point where the size of a hi-fi component is dictated more by the number of controls needed on the front panel than by the circuitry that has to go inside. Our fingers and hands have not been miniaturized like electronic circuits and you still have to be able to grab that volume control or tone control with your thumb and forefinger.

By the beginning of the 1970s the hi-fi component industry had settled into a comfortable pattern of growth. The youth of the world had been turned on to good sound; in the United States, next to owning a car or a motorbike, acquiring a good sound system became the most important single thing to millions of young people. Just about then another "breakthrough" occurred. As of this writing it promises to advance the quality of sound reproduction one more step in the direction of total musical fidelity.

How Many Channels Are Enough?

Early claims made for stereo sound reproduction stated that it provided "three-dimensional sound." If you think about it, though, the best that a stereo system can do is to provide "two-dimensional" sound in a living room. As we've mentioned before, in a live concert we hear not only the sounds coming from the performers and their instruments but from the concert hall as well. Call it the "acoustics" of the hall, the "ambience" of the hall or "reverberance," but the fact is that unless you can get that feeling of sound from everywhere around you, the system still falls somewhat short of the real thing.

Four-channel—or *quadraphonic*—sound was first introduced to the home listening public through the tape-recording medium, where it's relatively simple to record four tracks onto a strip of tape instead of two. Actually, multi-channel (more than two) sound recording had been used for years in recording studios where each performer was recorded on a separate track and subsequently "mixed down" to the final two-track stereo or single-track mono product. By mixing down to four tracks it becomes possible to "place" sounds anywhere in the listening room, if speakers are positioned in the four corners of the room. That's what the new four-channel systems are all about. The development of four-channel recording techniques for discs suggests that this major innovation will catch on in the coming years in much the same way that stereo discs and equipment took off in the 1960s. We'll get into the four-channel thing later on in this book.

I'm often asked what is likely to come next. Are we headed for eight channels and eight-speaker systems in a listening room of the future or will we someday go for sixteen? All I can say at this point is that four channels do permit the recording artist (and the listener) to position sounds anywhere in the listening room. While purists say that the back channels should be used only to provide a sense of "hall ambience," performers of rock and pop music were quick to sense that surrounding the listener with sound creates a sense of involvement in the music that just can't be achieved in any other way.

Quadraphonic sound can really be thought of as a new musical medium and, as such, there are no limitations or restrictions placed upon it. If the composer and performer feel that certain vocal lines should hit you from the rear, that's easily accomplished in four-channel sound. If the composer wants the listener to be onstage, surrounded by the performers, that too becomes easy with quadraphonic recordings. In short, four channels would seem to be enough to achieve all these goals. The only purpose that might be served by additional channels would be that of more precisely defining the location of sounds, much as fine-grain film can give better definition to a photograph than coarse-grain film. That seems to be a relatively small improvement compared to the extra cost that would be involved in doubling the number of sound channels once again, so I suspect that quadraphonic sound will be with us for a long time.

Stereo is not dead yet. The cost of a really good four-channel system is still quite high, compared to good stereo. If you're working with a limited budget a good stereo system that provides adequate sound power and low distortion is better than an inferior quadraphonic system that can't be upgraded later on. Just about every stereo system in the component category is so built that it can be easily converted to four-channel use at a later date. That, as we have said, is one of the nice things about the component approach.

As we begin to consider the things you ought to know about high fidelity components in ensuing chapters, we'll confine the discussion to stereo to keep things clear. Bear in mind, however, that when we talk about performance specifications of a speaker, a tuner, an amplifier or a record player, all of the important things to watch out for apply equally well to quadraphonic systems. Just adding additional sound channels does not replace the need for good performance in *each* of those channels.

We've come a long way from the time when putting together a good hi-fi component system required the equivalent of an engineering degree. Once you've selected the components you want (the hardest part of the job), gotten them home and taken them out of the carton, the system assembles in less than five minutes. The good sound lasts for years and years. So let's get on with the important business of how to choose the components for your dream hi-fi system.

Chapter 3

PUTTING TOGETHER THE ELECTRONIC COMPONENTS

As you begin to look into the matter of hi-fi components, you get the feeling that there are an infinite number of ways to put together a system. Looking at spec sheets you'll see complete receivers, amplifiers that have a lot of controls on the front panel and other amplifiers that don't have any controls at all. You'll see things called preamplifiers that look for all the world like some of the amplifiers but require amplifiers to work with them. You'll see components called tuners that look like receivers but aren't; and you haven't even considered such things as record players, tape recorders and—most important of all—loudspeakers.

To sort all this out, let's concentrate on the purely electronic components of a hi-fi system. There are really only three sections to this electronics. There's the *tuner* section, which picks up FM or AM radio signals sent out by radio stations and changes these inaudible high-frequency signals to audio signals of much lower frequency—which are still inaudible but correspond to the program information transmitted by the radio station. There is a *preamplifier-control* section that further amplifies program signals (whether they be those recovered from radio transmissions, record discs or magnetic tape record-

Fig. 3-1 A typical all-in-one stereo receiver.

ings) and allows you to change their amplitude, or level, and their tonal content. The preamplifier-control section also serves as your own switchboard, enabling you to select any one of the different program sources you may have connected to it. Finally, there's a *power amplifier* section that further amplifies these program signals and converts them to the energy necessary to activate your loudspeakers—which finally reproduce the actual sounds you want to hear.

The Integrated Stereo Receiver

By far the most popular way in which these three sections are put together is the all-in-one receiver. Such units contain all three sections just mentioned—the tuner, the preamplifier and the power amplifier—on one single piece of equipment. A typical stereo receiver is shown in Fig. 3-1. If all you wanted to hear was FM or AM radio, all you'd need to add to it would be a pair of loudspeakers and some form of antenna (for FM reception). Then you'd be all set the moment you plugged it into the wall.

In the early days of hi-fi the all-in-one receiver was looked upon by the experts as a poor man's compromise hi-fi component. In those days all the electronics of hi-fi used vacuum tubes that produced a lot of heat. If a designer tried to build an all-in-one receiver of high power capability (anything over about 10 or 20 watts per audio channel) he had to use large tubes which gave off even more heat. Such overheating tended to damage the more delicate components used in the tuner and preamplifier-control sections of the receiver. For that reason, early hi-fi enthusiasts usually bought separate power amplifiers (which were great for heating cold rooms on winter nights), separate preamplifier-control chassis and separate FM or FM/AM tuners.

In the late 1950s and early 1960s the hi-fi component industry "discovered" the transistor, or *solid state circuitry,* which was

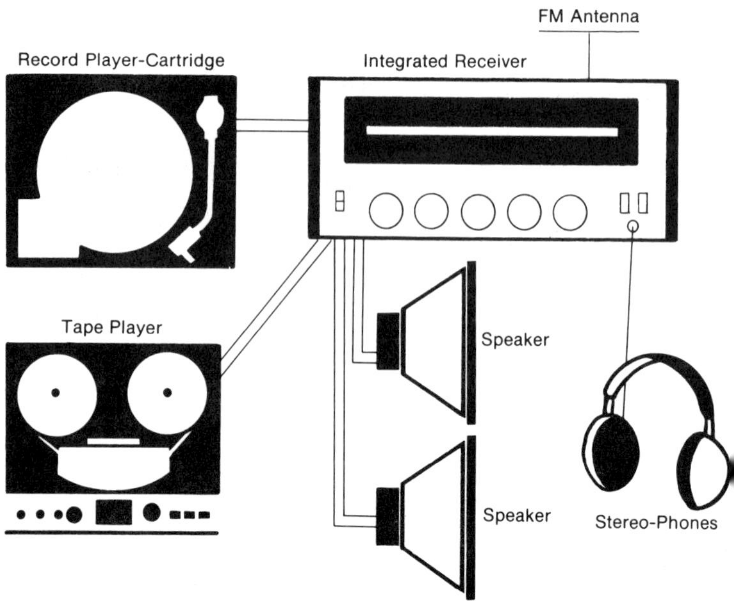

Fig. 3-2 A stereo system based on an integrated receiver.

already in use in millions of portable low-fi radios all over the world. Transistors are much smaller than tubes and produce far less heat than their predecessors. That's not to imply that transistors don't give off *any* heat. They do, and when a big solid-state power amplifier is pumping power to a loudspeaker, you can still feel a fair degree of warmth coming from the back of the amplifier, but it's just a small fraction of the wasteful heat produced by vacuum tube type amplifiers called upon to produce the same power.

As a result of the transition to solid state equipment, it became possible for designers to build all-in-one receiver chassis with power output capabilities as high as 50 or even 75 watts per channel—a figure great enough for most hi-fi system applications. We'll get into just how much power you need in a good system later on; for the present you can see that this new-found technology accounts for the increased popularity of the all-in-one receiver approach to hi-fi.

It figures that when all three sections of electronics are incorporated on a single chassis, certain cost savings result. There's only one physical chassis to deal with: one front panel and one cabinet or enclosure. Internally, only one power transformer and power supply is needed. Other parts (switches, controls and the like)

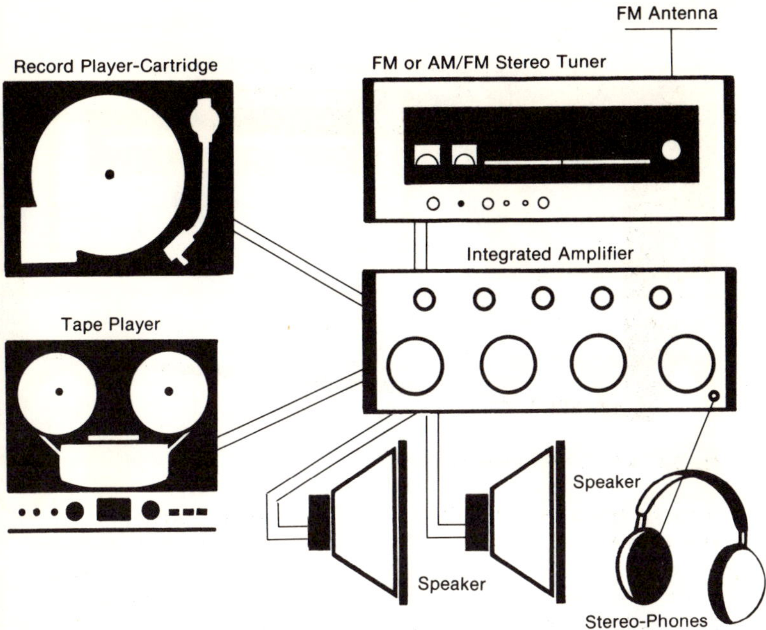

Fig. 3-3 A stereo system based on separate tuner and integrated amplifier.

don't have to be duplicated as they would be if separate units were used for each electronic section. For these reasons, the all-in-one integrated stereo receiver offers the greatest dollar value of all the system possibilities and probably accounts for from eighty to eighty-five percent of all sales of hi-fi electronics.

The diagram of Fig. 3-2 shows how all the elements of a complete hi-fi system go together using the integrated receiver as the "heart" of the system. Besides the pair of loudspeakers, there is a record player with its associated pickup cartridge and any one of several forms of tape recorders. A few receivers even have input facilities for connecting a microphone. Almost without exception you can even connect the audio signals from a TV set so that your TV sound comes out of your hi-fi speakers instead of from the very poor quality speaker usually supplied with even the best TV sets. Prices of decent stereo receivers start at just under the $200 figure and can run as high as $600 to $700, depending primarily upon power output capability and secondarily upon the control features and other performance specifications which we'll get into presently.

Fig. 3-4 A typical integrated amplifier.

Separate Integrated Amplifiers and Tuners

The next most popular way to put together the electronics of a component hi-fi system is illustrated in the diagram of Fig. 3-3. Here, the electronics has been divided into two separate components. The integrated amplifier now serves as the heart of the system and a separate tuner is shown connected to it. All the other components remain the same as before. A typical high-quality integrated amplifier is pictured in Fig. 3-4; a separate FM/AM stereo tuner of recent design is shown in Fig. 3-5.

In view of the economic advantages of the all-in-one integrated receiver, why would anyone choose this two-piece electronic set-up? I can think of

Fig. 3-5 A separate AM/FM stereo tuner.

two good reasons. First, suppose you have a limited budget to start with and your main interest is good record-playing facilities. Let's say that $500 is what you're starting with and you want the best sounding record reproduction you can get for it. You could probably find an all-in-one receiver for under $300, which would leave only $200 for a pair of speakers and a record changer. That isn't very much and could lead to an unbalanced system from a quality point of view. On the other hand, you might find an integrated amplifier for, say, $200 which has the same power output capability as the more expensive receiver and may even have more control features. That would leave $300 available for speakers and record changer and could make for a much better sounding system if all you care about for the moment is good record playing equipment. At a later date, when you've scraped together more money, you can always add a stereo FM or FM/AM tuner and connect it to one of the input pairs on your integrated amplifier. Taking it in easy stages, you would end up with a better system than if you had plunked your money down for an all-in-one receiver even though the *total* cost of your system may be a bit higher over the long term.

Another reason for choosing the separate amp/separate tuner route might have to do with basic performance specifications and extra-features availability. The only way in which makers of separate amplifiers can justify their place in the market is to design into them better performance and more control features. An example of more features might be the provision of two phono input jack pairs instead of the usual single pair. You might like to play records for casual listening using a record changer. Thus you can stack six or more records on a spindle and forget about them for several hours. For more serious listening, you might prefer to use a single-play turntable and tone arm combination and, perhaps, a better cartridge that can track the grooves of these records with less tracking force applied to the record groove. Having dual phono input facilities on an integrated amplifier enables you to connect a record changer to one pair of inputs and a manual turntable/tone arm combination to the second pair.

Integrated amplifiers generally offer more front-panel controls too. There may be better-acting tone controls (or more of them), superior rumble and scratch filters (for eliminating turntable low frequency rumble and surface noise from older records) and greater switching facilities than are found in typical all-in-one receivers. Finally, integrated amplifiers in a given price class usually offer higher output power capability and lower distortion performance than all-in-one receivers in the same price class. Of course, you've got to watch out

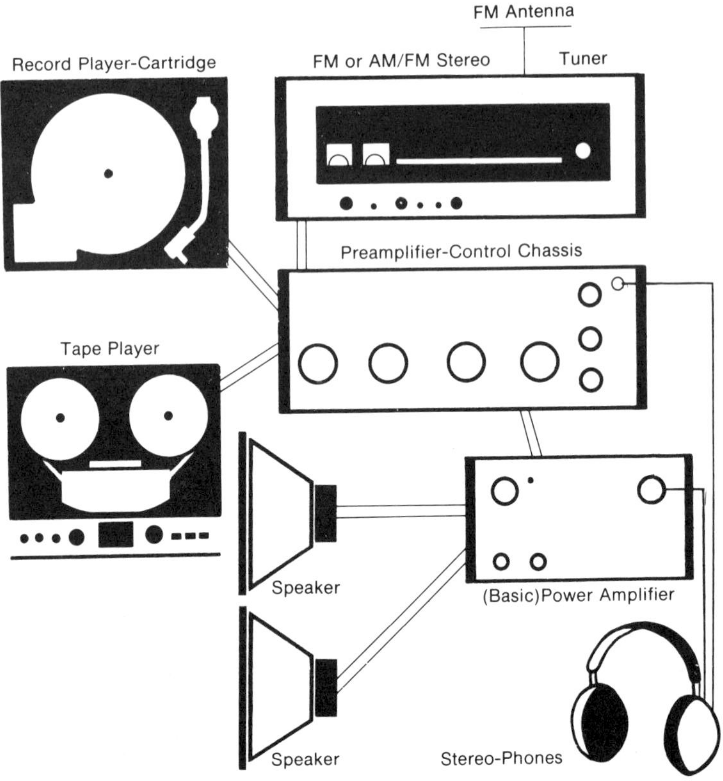

Fig. 3-6 A stereo system based on separate tuner, preamplifier and basic amplifier.

for that word "generally." There are always exceptions and contradictions. That's why a good understanding of performance specifications is so important and why several chapters of this book are devoted to the interpretation and understanding of those technical performance specs.

Separate Tuner, Preamplifier and Power Amplifier

Finally we come full circle in the electronic options of a hi-fi system: shown in the diagram of Fig. 3-6 is the most elaborate of systems, consisting of a separate tuner, a separate preamplifier-control chassis and a separate power amplifier, also referred to as a *basic amplifier*. The photo in Fig. 3-7 shows a modern stereo preamplifier while a powerful basic stereo amplifier is shown in

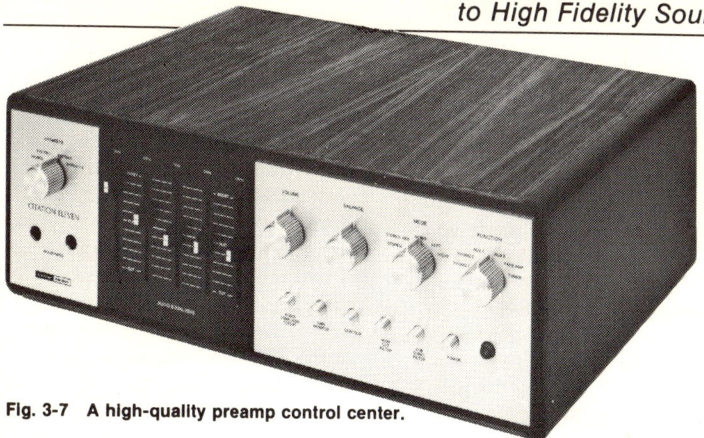

Fig. 3-7 A high-quality preamp control center.

Fig. 3-8. Although the units shown are all solid state and all of them are intended for stereo use, in all other respects the setup in Fig. 3-6 is really no different from the hi-fi system arrangements used in the very beginning by the earliest hobbyists.

The reasons for such an elaborate system are quite different, however. Basically, this type of system offers greater power output capabilities than either of the two arrangements shown earlier. We said that receivers can now be built with power outputs of as high as 50 to 75 watts per channel. Integrated amplifiers can be designed to produce somewhat higher powers than that. With basic power amplifiers there seems to be no limit, and in recent years we have seen power amplifiers having 150, 200 and even 350 watts-per-channel capability sold for home use. Admittedly, today's music buffs like their music loud, and playing loud means more amplifier power is needed. But isn't 350 watts bordering on the ridiculous? Not really,

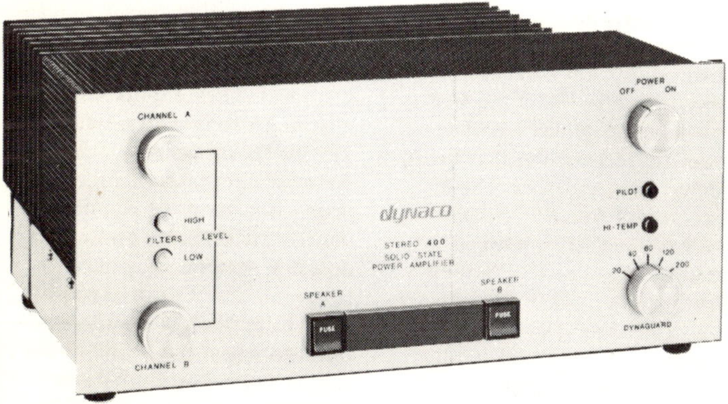

Fig. 3-8 A basic dual power amplifier.

as you will see when we talk about loudspeaker efficiency in the next chapter. There are some loudspeaker systems made which actually require that kind of power in order to produce the loud sound levels demanded by some listeners. Even if the speakers you choose don't soak up that much power there is always the possibility that you may want to equip several listening rooms with stereo speaker pairs and operate some or all of them at once from the same electronics. If you drive two pairs of speakers at once from the same amplifier, power is equally divided between the two pairs. If you use a 100 watt-per-channel amplifier for this purpose only 50 watts of power will be available for each of the two speakers used on that particular (left or right) stereo channel. If you have speakers located in three rooms only 33 1/3 watts will be available for each, and so on. You can be sure, too, that a manufacturer who sets out to build this kind of super-powered and super-cost amplifier is going to shoot for absolutely the best performance specifications he can devise, recognizing that his potential customer seeks the ultimate in performance. Some of these amplifiers are equipped with power level meters, power indicator lights and other refinements rarely found on the all-in-one integrated amps and hardly ever included in integrated receivers.

The separate preamplifiers used with these big power amplifiers are usually engineered with extra features as well. They may have level-indicating meters, microphone mixing facilities, extra inputs and outputs of all kinds, provisions for front panel connection of tape recorders for dubbing or copying tapes from one machine to another, multiple tone controls, better filters, better hum and noise characteristics and very low distortion figures. Again, we're speaking generally and are quick to admit that you may find low-cost preamplifiers on the market that don't have all these features and you may find high-priced integrated receivers available that have many of these same features in the single piece of equipment. These are the exceptions rather than the rule, and in choosing which of the three kinds of electronic systems you want, the advantages of added flexibility, power, etc. must be weighed against cost.

It goes without saying that the three-piece approach is the most expensive of all, if all other things are equal. I don't think there's anything wrong with holding out for this "best-of-all-possible-systems" approach if you can justify it in terms of genuinely desired performance and power needs. What does bug me is the large number of ultra-expensive three-piece electronic systems I run into where the proud owner neither needs nor appreciates these super refinements and has gone all the way for snob appeal or because some smart salesman conned him into such a system. My own recommendation would be to list the features—power requirements and such—that you think you need and then see if you can't find it in a complete receiver. If not, shop around for separate integrated amplifiers. If none of them fills the bill, look into the rarefied atmosphere of the separate amp, preamp and tuner combinations. In any event, don't start any of this until you've read the chapter on speaker selection; much of the decision-making process about electronics is directly related to your choice of speaker systems.

Other Hi-Fi Program Sources

Besides FM radio (and possibly AM radio, which I don't consider to be a true hi-fi program source) you'll want to equip your high fidelity component system with record-playing facilities. If you are anxious about record-changers and can't wait till we cover them, turn to Chapter 7. If you're big on tape recorders, you'll find them discussed in Chapter 8. If you've already dashed right out and bought the whole works and are anxious to get going (and too lazy to read the manufacturer's instruction booklets), turn to Chapter 9 for some tips on hooking together the system. Otherwise stay with it as we go on to examine the most critical components in your sound set-up: the loudspeakers.

Chapter 4

ALL YOU REALLY HEAR ARE THE LOUDSPEAKERS

Reading some of the hi-fi ads that appear in various publications can make you lose sight of a very basic truth. Brand X tuner is said to "sound" great, while Brand Y's amplifier produces such realistic "sound" that you'll swear you're hearing the real live thing! Well, I've got big news for all you ad writers and readers: tuners and amplifiers (or tape machines or phono cartridges, for that matter) don't make any sounds at all. The only element in a hi-fi unit that produces sound is your loudspeaker system.

This popular James B. Lansing L-100 uses bass-reflex principle.

Unfortunately, while the electronic parts of a system continue to come closer to absolute perfection, loudspeakers continue to represent the so-called "weakest link" in the hi-fi chain. If loudspeakers were even close to perfection, they'd all begin to sound alike. The fact that they don't means that each is contributing some form of audible *distortion,* to use the very harshest term around. Distortion, in sound reproduction, means any deviation from the original sound. It isn't limited to harmonic distortion or intermodulation distortion, although those are among the most familiar kinds.

Amplifiers and tuners of better quality consistently produce output signals containing well under one percent distortion. Some of these products are so distortion-free that it has become increasingly difficult to measure the small amounts of distortion they do produce—the measuring equipment just isn't up to the task. Not so with speakers. When trying to reproduce low-frequency bass tones at high sound levels, distortion can reach figures of five percent, ten percent, even fifteen percent. Somehow we listeners are able to tolerate such sounds.

I don't have any actual statistics from the United States Patent Office, but I'll bet they get dozens of patent applications every year—each one claiming to be the sought-after breakthrough in speaker design.

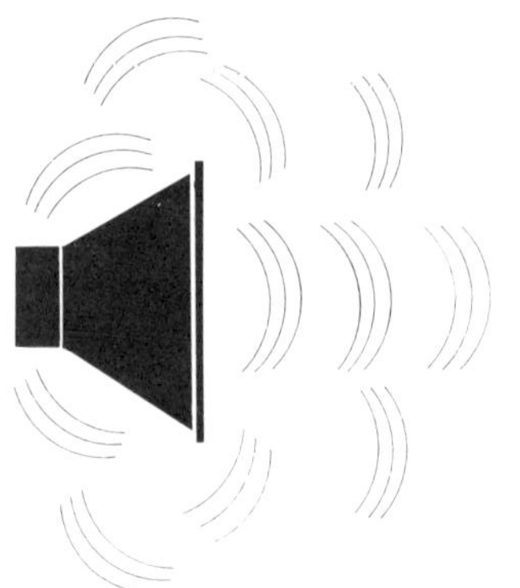

Fig. 4-1 Cancellation of sound waves in unbaffled speaker.

Speaker systems are generally divided into five or six major categories. An understanding of the basic principles used may prove helpful as you shop for *the* speaker system that sounds best to your ears.

First it's important to understand that a raw loudspeaker trying to reproduce music fights a losing battle. Sound is produced by the cone of a loudspeaker pushing air into the listening room, but air is pushed in front of, as well as behind, the speaker cone. Thus, if both front and back of a speaker are exposed, the front is *pushing* air while the back is *pulling* air. The two effects tend to cancel each other, especially at low frequencies. This idea is shown dramatically in Fig. 4-1. The basic idea behind enclosing a speaker in a box is to seal off the back air compressions so that they don't interact with—and cancel—the front air compressions. Properly enclosed, the speakers begin to sound great—yet still different from one another.

Infinite Baffle Enclosures

Mounting a speaker right into a wall is one way of isolating the back waves from the front waves. When hi-fi component assembly was more of a do-it-yourself affair, many hobbyists simply bought raw speaker elements and mounted them through walls, closet doors and even through main structural walls of their homes with the backs protruding into the garage.

Applying this concept to cabinetry meant designing large enclosures whose cubic volume of air was great enough not to restrict the motion of the speaker cone. These were the so-called *infinite baffle* enclosures that were popular in the monophonic '50s. They were *large,* often ten cubic feet or more. With the coming of stereo in the '60s their popularity waned.

Bass Reflex Enclosures

By cutting a proper-sized hole or "port" in an enclosure it was found that the back-of-the-cone sound wave would reinforce the front sound wave—at least over a small range of low frequencies—and the so-called *bass reflex* enclosures were created. Somewhat smaller in size than equivalent infinite baffle designs, these speakers offered a partial solution to the space problem as stereo became increasingly popular. You'll still find a good many speaker products that employ

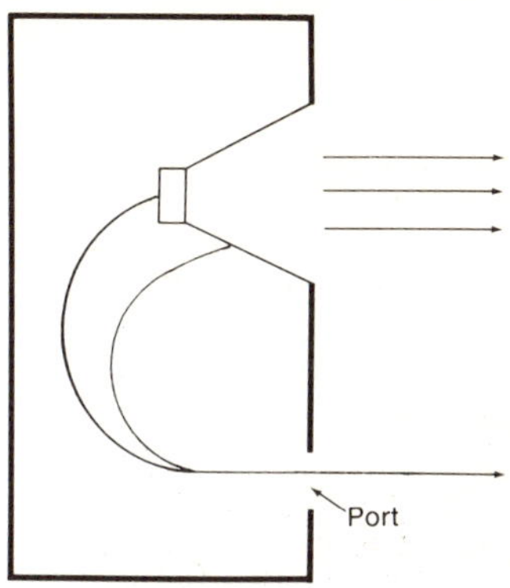

Fig. 4-2 Bass-reflex speaker enclosure.

this principle. If properly designed they can be quite efficient (requiring few amplifier watts) and reasonably flat in frequency response. If improperly designed they can sound like a tubby open-backed console, with its characteristic "one-note" bass boom that seems to attract the unsophisticated listener who runs his system with bass and treble controls turned all the way up. A cross-section diagram of the typical bass reflex is shown in Fig. 4-2.

Horn Systems

For maximum speaker efficiency and deep authentic bass response nothing beats the *horn* enclosures. Based upon sound acoustic principles, the horn enclosure provides an excellent match between the speaker element and the surrounding air environment. You've seen this type of speaker system in large stadiums and at outdoor concerts. In its pure form the horn has to be several feet long to do a good

Klipschorns need room corners—and lots of space.

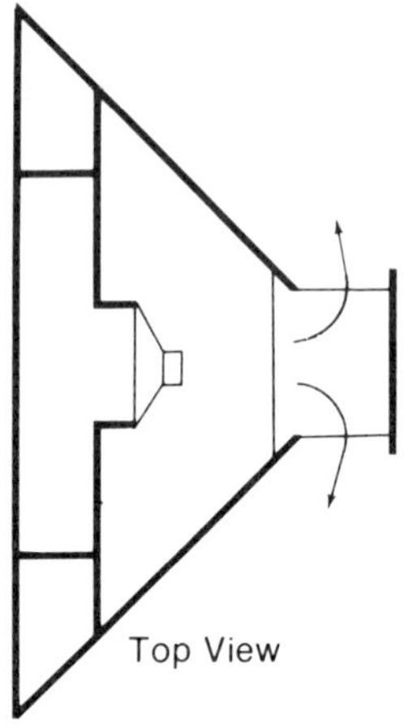

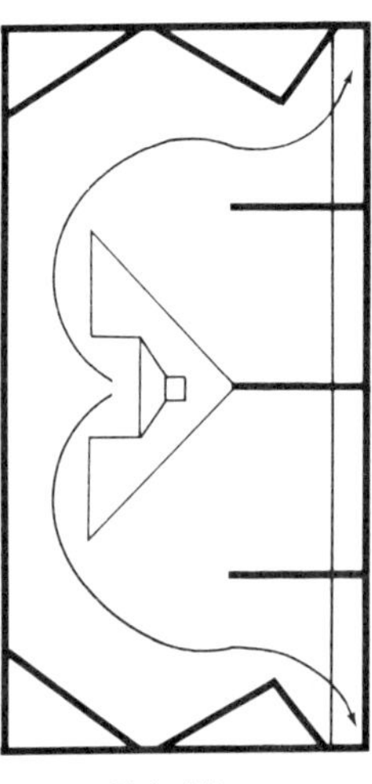

Fig. 4-3 Rear-loaded horn—Klipschorn enclosure.

job. By "folding the horn back on itself" designers have come up with home units that are of reasonable (though large) proportions and still take advantage of the horn principle.

Perhaps the best-known of these is the famous Klipschorn, invented by Paul Klipsch of Hope, Arkansas. Mr. Klipsch has been turning out pretty much the same speaker enclosures for nearly two decades. He still insists that the all-new, popular "bookshelf" designs are a flash-in-the-pan and that only horn enclosures can provide natural-sounding bass. The fact that he's been able to make a good living from selling his monsters attests to the number of Klipsch fans who wouldn't be caught dead with a power-hungry, low-efficiency air-suspension speaker system in their listening rooms. Generally, horn-type enclosures do even better when they are positioned in the corners of a room; the floor and the two adjacent walls actually act to extend the dimensions of the horn so that the room itself becomes part of the overall acoustic design. Cross-sectional views of the Klipschorn are shown in Fig. 4-3.

Acoustic Suspension Systems

By far the most popular systems in use today are the so-called *air suspension* or *acoustic suspension systems.* First marketed by Acoustic Research (AR), they represented a real breakthrough in speaker enclosure design. They hit the scene at just the right time—when the dual speaker needs of stereo were becoming a big problem for the home owner and apartment dweller.

Smaller in size than other enclosure types, they operate on a completely different principle. Most speaker element cones are rather stiffly suspended from the metal framework of the loudspeakers. If you push the cone with the fingers and let go, it snaps back to its neutral position in a quick, springy way. If you were to remove the front baffle of a typical bookshelf design and poke the woofer cone you would find that it can be displaced from its center position by a greater amount. In returning, it has a kind of "cushy" feeling as it meanders back to its neutral point. Examined in the raw, speaker elements that are used in air suspension systems have their cones very loosely suspended. Operated in free air they don't sound like much, but when sealed into a box of relatively small dimensions the bottled-up volume of air behind the cone acts as part of the suspension, tightening up the cone action. In other words, the amount of air inside the box is critically defined—unlike the case of large infinite baffle designs in which the greater the volume the better.

Designs of this type can produce very creditable bass response. Because of their small size, they are a natural for stereo and even more of a find for four-channel set-ups, in which four speaker systems have to be used. Of course you can't get something for nothing in speaker design. The thing that these speakers give up is *efficiency.* Most air suspension speaker designs require lots of power for a given amount of sound volume—often ten or twenty times as much as is required by larger, more efficient speaker system types.

That's probably why so many amplifier manufacturers are getting into bigger and bigger power amplifier wattages. Years ago we would have thought it a big put-on if someone offered a 100-watt amplifier for sale. Today such power capabilities are commonplace and 300- and 400-watt units are offered as well.

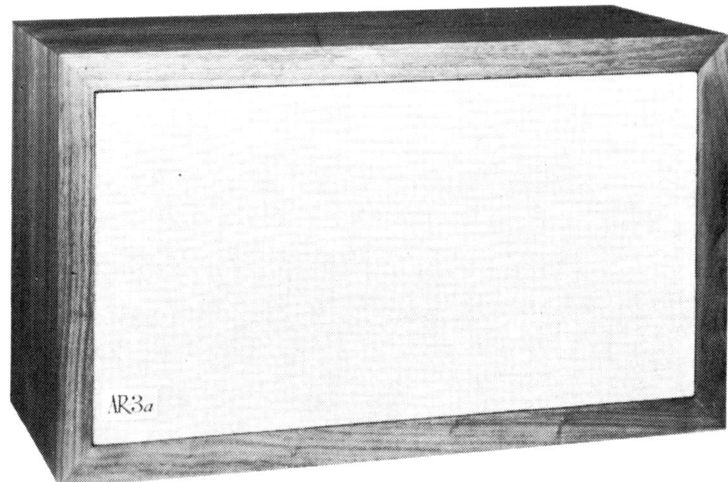

A pair of well-accepted acoustic suspension designs from Acoustic-Research (AR) and . . .

When we speak of low-efficiency speaker systems, we don't mean low quality. Efficiency is merely an indication of how many watts of electrical power you have to feed a given loudspeaker system to extract a given amount of sound power. In the case of air suspension systems that's usually a big number. That's why the rush seems to be on for high-powered amplifiers.

While bookshelf types dominate the field, there is a strong tendency towards larger, more efficient systems. Many manufacturers are thumbing through old dusty designs and coming up with floor-standing models once again. Often new approaches are used with regard to decor. We've seen speaker systems disguised as marble-topped end tables, bongo drums, futuristic plastic spheres and you

. . . more recently, Advent Corporation.

name it—all created in the name of furniture styling. Some of these really come off and blend well with the other contents of a living room. Some, unfortunately, are just too pushy and far-out to have gained wide acceptance so far. One thing's certain: there are probably more individual speaker systems available, made by more companies, than there is any other single component in hi-fi land. It all gets back to that first premise: no one of them is perfect. Any gadget that is that far from perfection is going to have lots of variations, with each claiming to surpass the next in performance. It figures, then, that speaker auditioning ought to take up the major part of your shopping time.

Apart from the Bass

Most of the differences in speaker design we've talked about so far have to do with bass response. For the most part, enclosure size and style have very little to do with the reproduction of highs and ultra-highs. In considering high frequency reproduction, the real differences in design philosophy have to do with the *dispersion,* or directionality, of those higher frequencies. Here there are two approaches: the *direct-radiating* format and the *omni-directional* school.

Most high frequency tweeters are mounted on the front surface of the speaker system. Highs, though, are very directional in nature and radiate in a narrow beam. If you've ever been bothered by a neighbor's hi-fi system you've probably noticed that all you hear is the dull, thud, bass quality of the music, because the highs just don't go around corners or through walls. In the case of direct-radiator designs, speaker makers try to build tweeters with as much angular dispersion characteristic as possible. Often more than one tweeter will be used, each positioned at a slightly different angle to provide the widest possible dispersion.

A logical extension of this leads to the omni-directional design in which speakers are mounted away from the listener or even pointing straight up towards the ceiling. The idea here is for the high frequencies to take advantage of reflecting surfaces and bounce back at the listener from several different directions. This is supposed to result in a smooth noncritical distribution of treble sounds around the listening room.

Serious audiophiles are about evenly divided on the validity of this approach. That's why you'll find a variety of modern speaker systems that subscribe to this theory and an equally wide assortment of models that don't. The argument becomes even more heated when you consider quadraphonic sound. If pinpointed directionality from the four speaker locations is the object, you'd conclude that directional highs and mid-range tones would be desirable. Yet many listeners who have begun to experiment with

Bose system uses direct/reflecting sound principle preferred by some listeners.

four-channel sound are just as positive that omni-directionality is even more important when it comes to four-channel listening.

Choosing the Right Loudspeaker

Before even considering *how* to choose the right loudspeaker for your hi-fi system, a more fundamental question might be *when* to choose that loudspeaker system. The answer is: *before* you choose any other component of your high fidelity set-up. While this may seem to contradict some of the advice you've gotten from other sources, you can see that with this matter of widely varying efficiencies, choosing the right loudspeaker and knowing its power requirements and capabilities will help you in choosing the rest of the electronics of your system.

As we've said, most of the popular bookshelf or air-suspension systems used today are highly inefficient. As an example, their efficiency might run anywhere from ½ of 1 percent to 2 percent. Typically, then, a bookshelf speaker having an efficiency of 1 percent might require 100 watts of audio electrical power input to produce 1 acoustic watt of sound power output. At the other extreme, highly efficient floor-standing speakers may have efficiencies as great as 10, or even 15 percent. Considering the horn enclosure type of system, for example, that has 10 percent efficiency, it could be expected to produce the same 1 acoustic watt of sound power output with an electrical amplifier power input of only 10 watts, a difference of 10-to-1 compared with the inefficient air-suspension type given in the earlier example.

Obviously this business of efficiency has a greater effect on your amplifier power requirements than any other single thing in your acoustic environment. Many people have said that larger listening rooms require greater amplifier power, and that heavily absorptive rooms (those having carpeting, draperies or other sound-absorbing materials as part of their furnishings) also require greater amplifier power to drive speakers to adequate sound levels. You can now understand, however, that all of these acoustic factors are really far less important than the efficiency of the loudspeaker you choose. If space is an important consideration in your listening room, you'll undoubtedly favor one of the smaller bookshelf types. In doing so you'd better recognize the fact that you're going to need more amplifier power (and therefore a more expensive amplifier) to drive that type of loudspeaker system.

Speaker Power Requirements

Most speaker manufacturers generally recommend the minimum amount of power required to drive their loudspeaker systems to adequate listening levels. If a speaker manufacturer tells you that it will take 20 watts of power to drive his speaker, it would be a mistake to settle for anything less in the power capability of your amplifier (or the amplifier section of your all-in-one stereo receiver). Some manufacturers have also begun to quote maximum power handling capability of their loudspeaker products.

If anyone ever tells you that the loudspeaker in a high fidelity component system is indestructible, don't you believe it! It is not unusual to blow out speakers or damage them beyond repair by applying too much amplifier power to them. Furthermore, even the most liberal of warranties given by the speaker manufacturer will not cover misuse or abuse—such as applying too much power to the loudspeaker. As you zero in on your favorite-sounding speaker, you should be aware of both its minimum and maximum power requirements. Bear in mind that if you ever plan to connect additional loudspeakers to the same amplifier, doing so will cut the available power to each speaker.

Frequency Response

Most manufacturers don't even bother to publish curves of speaker *frequency response*. That is because there is no standard way in which to measure the actual frequency response of a loudspeaker and present it meaningfully. Trying to make accurate frequency response measurements in a typical listening room is quite difficult. The room itself often has reflective or reverberant characteristics that will distort or upset the frequency response curves obtained in this way.

As a result of this problem, what little information presented with regard to frequency response is usually vague and meaningless. Statements such as "Frequency response: 18 Hz to 17,000 Hz," made by a speaker manufacturer, don't mean much. All they tell you is that there is *some* output at all of those frequencies; they don't tell you how much the amplitude of that output varies across the audio spectrum covered. A look at Fig. 4-4 (a typical frequency response curve measured in a totally dead soundroom) will give you an idea of why speaker manufacturers are reluctant to publish such curves. If you ever saw a frequency response curve like that for an amplifier you would never even consider its purchase. In the case of speakers, that's about the kind of curve you can expect.

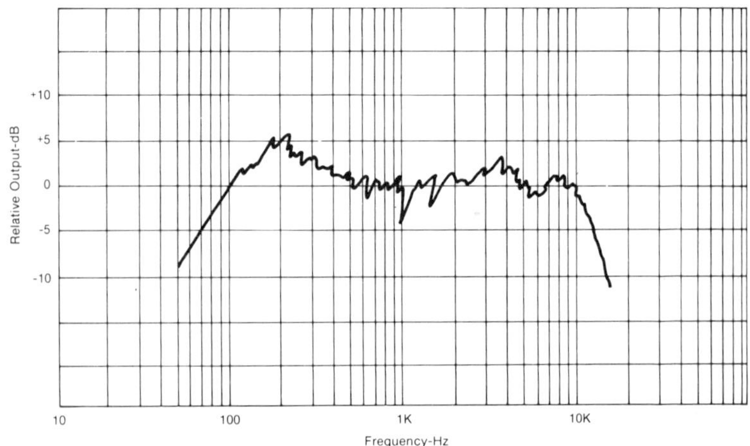

Fig. 4-4 The frequency response of most speaker systems is far from flat.

Those Mysterious dBs

Speaking of Fig. 4-4, the vertical axis of that graph is labelled "Relative Output—dB". Since we'll be talking about dBs, or *decibels* throughout the remainder of this book, this might be a good time to explain them. Decibels, unlike other quantitative measurements, have no magnitude of their own. That is, you can't say something measures 8 dB. You *can* say that something is 8 dB louder than something else or that a voltage is 6 dB greater than some other voltage; or that a power level is 3 dB greater than some other reference power level.

When it comes to sound and hearing, 1 dB is about the lowest change of sound level that most of us can hear. A change of 3 dB is audible to almost everyone as a significant change of sound level. A sound that is 10 dB louder than another will sound about twice as loud to listeners, even though the power ratio is then 10-to-1. That's because our hearing is not linear in its response but logarithmic. For this reason dBs are calculated on the basis of a logarithmic curve rather than a linear one. When something is twice as powerful as something else, it is "3 dB" greater in power. A change of 6 dB in power means something is four times as powerful as something else. A 12 dB change represents a 16-to-1 power ratio, etc.

Decibels can be used to define the relative levels of voltages and currents in electrical circuits too, but the formula for figuring them out is a little different. If you're interested in the actual math formula for dB calculations of either two voltage ratios or two power (or sound level) ratios, you'll find it spelled out in the Glossary at the back of the book.

Speaker Impedance

One speaker specification that may be of significance is the speaker's impedance. Usually speaker impedance will be rated by the manufacturer as nominally 4 ohms, 8 ohms or 16 ohms. Just about any modern solid-state audio amplifier can work well with speakers having any of these impedances. There is one thing to watch out for: if you plan to add speakers in other rooms at a future time, and if you plan for both the main speakers and the remote speakers to be operated simultaneously, it would be a good idea to avoid speaker impedances of 4 ohms. Amplifiers, when operating into impedances of *less* than 4 ohms, have a tendency to blow their fuses, destroy their expensive output transistors or blow themselves up. So you should try to choose a loudspeaker system having an impedance of 8 ohms or greater. When two 8-ohm speakers are connected in parallel, the resulting net impedance is 4 ohms—a value still safe for the majority of today's solid-state amplifiers or receivers.

Tips on Auditioning Speakers

When it comes right down to it, the only way to choose the loudspeaker that suits your needs and taste is by extensive listening. Choose an audio shop that has a wide assortment of popular speaker makes. Select a group of speaker models that fits in with your budget. First ask that any two of these be compared (preferably in stereo, which means two versus two) in what is called an *A-B test.* While you are listening to a familiar recording, the salesman should quickly switch from one pair of loudspeakers to the other.

As we have said, speakers vary widely from brand to brand and even from model to model in efficiency; some speakers fed with 5 or 10 watts will really blast you out of the room; others produce no more than background-music level when driven with the same 5 or 10 watts. The inefficient ones aren't "bad"; they simply require more power to produce a given amount of sound. A properly outfitted showroom will make due allowances for these differences in efficiency by correctly attenuating the "high efficiency" jobs, so that when they are compared with their less efficient competitors the loudness level is exactly the same when the switch is made from unit to unit.

Most of us can't honestly say which of two sounds is louder unless there is a difference between them of 2 or even 3 decibels. With lesser differences in level, we won't specifically relate the difference to loudness but tend to favor the higher output model on a quality basis. In other words, we gravitate towards the slightly louder-playing model without really being able to pin our preference down to the simple fact that it's playing at just a little louder level. The loudness factor should be eliminated from all comparison tests if at all possible.

Once you have selected the best of the two first pairs of speakers you listen to, compare it with a third stereo pair. If possible bring along your favorite recording or one with which you are very familiar. Believe it or not, attending one or more live concerts before you go shopping for speakers is a very good idea. Many of us are so accustomed to listening to "canned" sound, that we tend to forget what the real thing sounds like. Continue to make comparisons, two-by-two, of all the speakers in which you think you might be interested.

When you have finally selected your perfect speaker pair, ask the salesman to place a pair of them on the floor—if that's where you'll put them in your listening room; or on a shelf—if that's where they're to go. Listen to them for an extended period of time using several kinds of musical material. If you're still convinced that you've chosen the "dream" system, take them home.

Don't be surprised if when you set them up in your own listening room they sound a little bit different from the way they sounded in the audio showroom. Room acoustics have a great deal to do with the way speakers perform. No two rooms are alike acoustically. Some of the better audio stores will permit you to take home a pair of speakers on a trial basis and may even allow you to trade them in for another selection after a few days of listening.

Above all, try not to be swayed by the opinions of your "knowledgeable" audiophile friends. What sounds good to them may not necessarily sound good to you —and you're the one who has to listen to those speakers for years to come.

A final word of reassurance: remember that if after a few months you find that your listening skills have grown sharper and you really want more expensive or better-sounding speakers, you can always shift the first pair into your bedroom for hi-fi sound in two locations.

Chapter 5

FM TUNERS—HIGH FIDELITY RADIO

Whether you listen to *FM* radio using an all-in-one receiver or a separate tuner, the circuit features and specifications involved in this very important music program source are the same. Because FM (or AM, for that matter) involves radio broadcasting—unseen "waves" travelling many miles in practically no time at all—the average audiophile tends to place this medium in a class by itself. He usually avoids the seemingly incomprehensible specifications and descriptions of this product.

FM radio isn't all that complicated. An understanding of its workings, advantages and limitations should enable you to make an intelligent selection when it comes to buying the needed equipment to use for FM reception.

FM is a "line-of-sight" kind of radio transmission. The frequencies used for FM radio (88,000,000 Hz to 108,000,000 Hz—abbreviated 88 MHz and 108 MHz) can't "bend around the horizon" like lower-frequency (short wave) radio can do. Using a very sensitive FM tuner or receiver and a good outdoor antenna, you'll be lucky if you can pick up stations 50 or 60 miles away with reasonable clarity and noise-free performance. One of the virtues as-

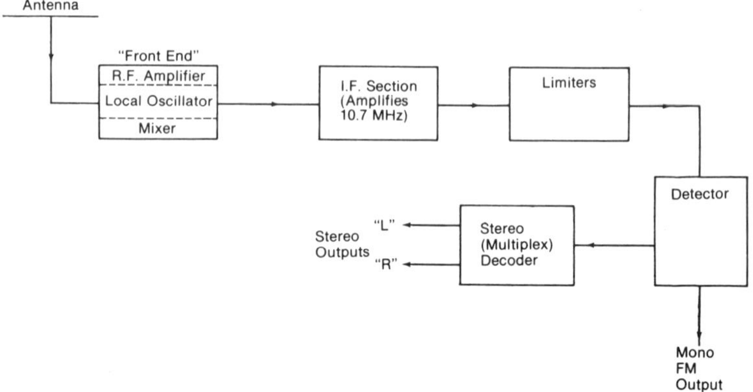

Fig. 5-1 Simplified block diagram of FM tuner.

cribed to FM radio is that it's "static free." Actually, its ability to reject static and noise is built into the tuner or receiver rather than into the FM broadcast system itself.

The block diagram of Fig. 5-1 shows the different sections that make up an FM tuner. The front-end picks up the radio signals intercepted by the antenna, amplifies them and converts them to an *IF (intermediate frequency)*—usually 10.7 MHz. The incoming radio signal actually varies in frequency whenever there's program material being broadcast. While the station's nominal frequency might be 101.1 MHz on the dial, the audio information being broadcast causes that frequency to vary up and down—from a low of 101.027 MHz to a high of 101.175 MHz. The *louder* the program the greater the departure from center frequency of 101.1. The frequency of the music or speech being broadcast determines how many times per second the station frequency varies above and below its nominal center point. This is illustrated in the diagram of Fig. 5-2.

The diagram also shows how the FM system differs from AM. In *AM,* the station frequency remains the same all the time; the *amplitude* of the radio waves changes in accordance with the audio program being transmitted.

To return to our block diagram: the IF signal is then amplified and limited. It's this limiting action that makes FM noise-free. Most noise and static is AM in nature. Therefore noise added to a radio signal will change its overall amplitude, as shown in Fig. 5-3. In FM, since we're only interested in detecting shifts of *frequency* (corresponding to the audio information we want to recover), we can slice off the noise from the top of the waveforms and not lose any

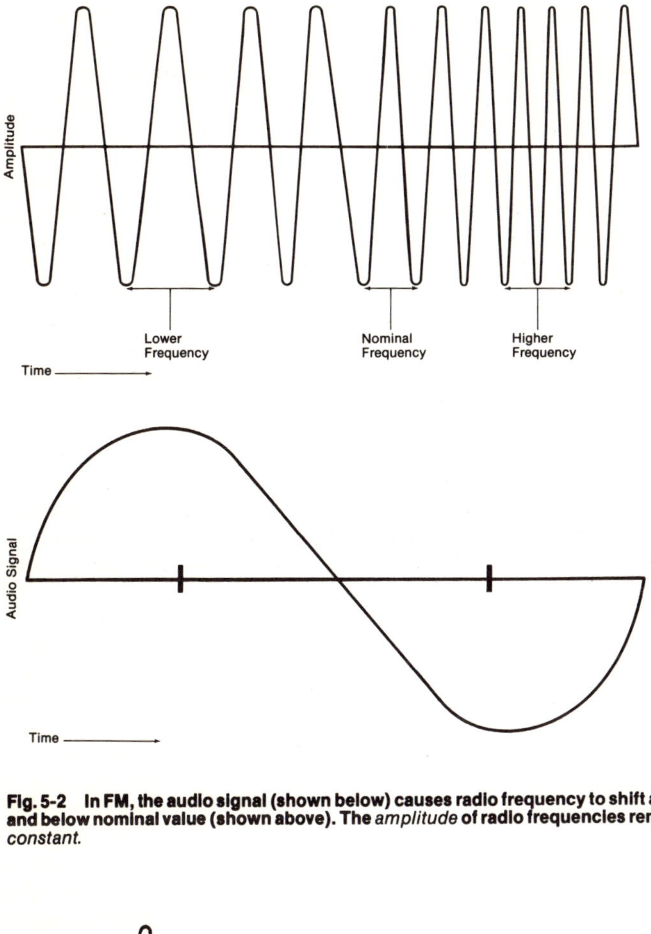

Amplitude

Lower Frequency

Nominal Frequency

Higher Frequency

Time

Audio Signal

Time

Fig. 5-2 In FM, the audio signal (shown below) causes radio frequency to shift above and below nominal value (shown above). The *amplitude* **of radio frequencies remains** *constant.*

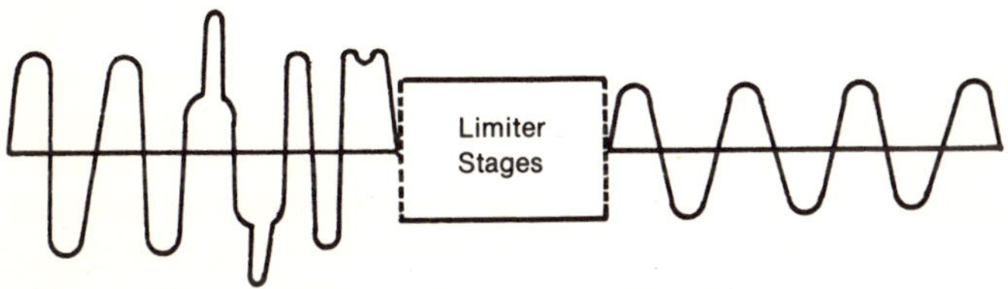

Fig. 5-3 Limiters chop off AM variations (static) of FM signals.

program content. That's just what the limiter does. Stripped of its external noise, the signal is then demodulated, i.e., the audio is recovered from the IF signal, by means of an FM detector. If we are dealing with a mono tuner that's all there is to it. For stereo tuners another circuit function has to be performed, that of stereo or multiplex decoding.

The Stereo Signal

In mono FM we can receive audio information from 30 Hz to 15,000 Hz. That top limit falls a bit short of the 20,000 Hz we said is needed for "real hi-fi," but it's high enough so that no one would call FM radio a low-fi medium. The FM station, however, can transmit higher frequencies within its allowed bandwidth (which is 200 kHz wide) without having them "spill over" into the next channel.

In the case of stereo broadcasting, that first band of frequencies from 30 Hz to 15,000 Hz is used to send out the *sum* of the left and right program channels (L+R). The two stereo channels are simply mixed together and sent out as one audio signal in the usual way. The owner of a mono FM tuner therefore hears the *total* program, not just a left-only or a right-only signal. As shown in Fig. 5-4, the *difference* between the left and right programs is transmitted via a modulated high-frequency sub-carrier, using mod-

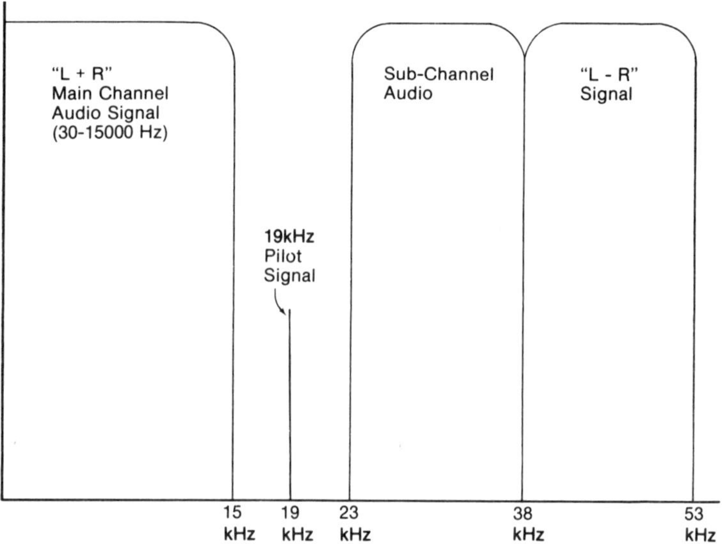

Fig. 5-4 Make-up of composite stereo signal. Pilot signal is sent along to help *lock* **38kHz sub-carrier in receivers.**

ulating frequencies ranging from 23 kHz to 53 kHz. These are way above the limits of human hearing. There are other fine points but the important thing to understand is the idea of a difference signal. If you'll admit that we can electrically *add* two audio signals together to make up the L+R signal, you should believe that it's possible to *subtract* one electrical signal from another to form a difference, or L-R program signal.

The L-R signal, hidden within those high superaudible frequencies recovered at the detector of the FM tuner, is not of itself audible. One job performed by the stereo decoder circuits of Fig. 5-1 is to translate those high frequencies back down to the audible range, i.e., to recover the L-R signal as real audio. Once that's done, the recovered L-R audio is added to the already audible L+R. (L+R) + (L-R) = 2L, if you remember your elementary algebra. That's 2L without any "R" mixed in—the original "L" program (you can disregard the coefficient "2"; that just means it's twice as loud). Then, if you subtract the recovered L-R signals from L + R you get (L + R) - (L-R) = 2R—the original "R" signal—with not a trace of "L" mixed in. Besides stereo, some stations also transmit private background music services paid for by subscribers and known as SCA programming. That's really all you need to know about how stereo FM works if you're considering the performance of a tuner you plan to buy.

Tuner Specifications

Sensitivity

When you pick up your first specification sheet or brochure concerned with FM tuner performance, don't panic. The terms and tabulated numbers may seem complicated but they're really not. Usually the first spec given is the *sensitivity,* sometimes called *IHF* sensitivity. (IHF stands for the Institute of High Fidelity which, among other things, sets up standards for measuring the performance of its members' products in order to universalize numerical meanings.) Sensitivity describes the ability of a tuner to receive very weak or distant signals and turn them into listenable audio. The number is usually given in *microvolts* (millionths of a volt, usually abbreviated uV) and, in theory at least, the lower the number the more sensitive the set.

Today, however, most better-quality tuners differ little as regards sensitivity; whether a spec reads 2.0 uV, 1.9 uV, 1.8 uV or even 2.1 uV is not terribly important. By definition, with this amount of signal strength reaching the tuner, the program (audio) level will only be about 30 dB greater than the background noise and distortion. Put another way, about 3 percent of what you hear will be either noise, distortion, or a combination of both. That doesn't make for very good listening but the sensitivity spec stated in this way is a throw-

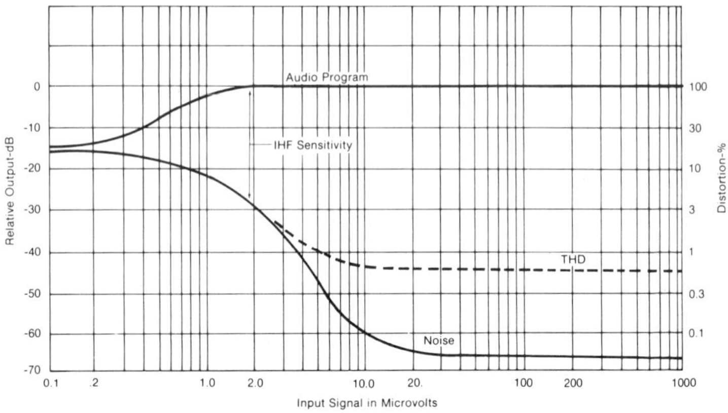

Fig. 5-5 Graph of typical noise and distortion in an FM tuner. IHF sensitivity is reached at 2.0 microvolts. Ultimate S/N is about 65 dB and ultimate THD is about 0.6 percent.

back to earlier days when a 30 dB signal-to-noise and distortion ratio was considered to be a "least usable signal" condition. I wouldn't call FM radio hi-fi until the noise is down about 50 dB below (0.3 percent of) the desired program material.

Signal-to-Noise Ratio

With further increases in signal strength, the noise and distortion does decrease, as shown in the curves of Fig. 5-5. Both reach a minimum and level off at more realistic signal strengths. The lowest noise figure is expressed as the *signal-to-noise ratio* in the spec sheets, or just simply *S/N.* A good S/N ratio in a modern tuner would be anything above 60 dB and I've seen some products that have S/N ratios as high as 70 or even 75 dB—which means the noise is just about inaudible if the signal strength is great enough.

Total Harmonic Distortion

The same graph shows that the distortion also tends to decrease with increasing signal strength until it reaches its ultimate or lowest value. In our example, this value is just under 1 percent, which I'd consider hi-fi quality. Even so I've measured distortion figures on better tuners down to 0.3 percent and even lower. You'll rarely find a spec sheet that also tells you how low the *total harmonic distortion* (sometimes abbreviated THD) is when receiving *stereo* broadcasts or, for that matter, how low the background noise (S/N) becomes when listening to a stereo broadcast. Some manufacturers have started to list these specs but most would rather not (so long as they don't have to) because the numbers usually look worse in stereo than in mono. Still, it's possible to find tuners with less than 1 percent THD even operating in the stereo mode.

Selectivity

Another important term you'll find in tuner spec sheets is *selectivity.* The value is given in dB and the higher the number the better. Selectivity describes a tuner's ability to tune in a given station without picking up interference from stations operating at nearby frequencies. Normally the FCC doesn't assign station frequencies in a given city closer than 800 kHz (four channel widths) apart. Suburban areas close to big cities may have stations at frequencies only 400 kHz away from metropolitan stations. For all this care in station frequency assignment, today's tuners are so sensitive that they can receive stations from widely separated locations with possible interference. The selectivity number stated in the spec sheet tells you how much stronger a signal removed in frequency by 400 kHz from the desired signal must be in order to interfere with the desired signal. A good selectivity figure would be 50 dB or better. Some sets claim really great figures of 70, 80 or even 90 dB of selectivity.

Capture Ratio

Capture ratio, another listed specification, works just opposite to selectivity. Here the lower the number (stated in dB), the better the spec. Capture ratio is concerned with the possibility of two stations operating on the very same frequency. Ordinarily this condition is highly unlikely unless you're located squarely between two distant large cities and can pick up different stations from each—both operating at the same assigned frequency. Indirectly, however, capture ratio also tells you something about the set's ability to reject signal reflections received from the same station. In Fig. 5-6 a station signal is received at a home FM antenna by the direct path shown and by reflection from a nearby metal structure such as a water tower or a large metal skyscraper. In TV sets this kind of thing results in "TV ghosts" or multiple images on the TV screen. In FM (and more noticeably when listening to stereo FM) the effect is called *multipath reception.* In really severe cases it can cause audible distortion, increased background noise and even loss of separation in stereo listening.

You might think of the direct and reflected signals as two "stations" operating at the same frequency. The capture ratio tells you how much stronger the desired signal must be than the undesired signal in order for the tuner to reject the latter. A typical capture ratio figure for a good set will be 3.0 dB or lower. About the best I've ever measured is just under 1.0 dB.

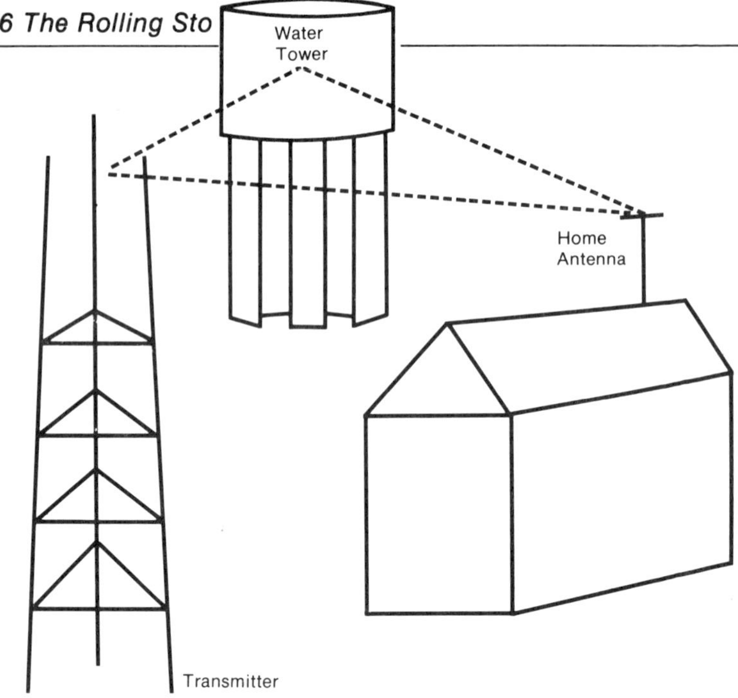

Fig. 5-6 How reflected signals can cause multipath interference in FM and stereo FM reception.

Other Tuner Specs

Less important monophonic specs you'll see detailed on the tuner or receiver manufacturer's literature are such things as *image rejection, IF rejection* and *spurious response rejection.* These three have to do with the ability of the tuner to reject signals that sometimes pop up in the middle of the FM dial where they don't belong. A good example of poor image rejection is the voice of a commercial airline pilot or control tower suddenly interfering with an FM program and coming in loud and clear. Good numbers for all three forms of extraneous signal rejection are anything above 60 dB. Some modern tuners do as well as 80 or even 100 dB in these characteristics.

While manufacturers tend to tell you little about the performance of their FM products in stereo (most of the numbers come out poorer than in mono), almost all of them will boast about the stereo separation capability of their product. *Separation,* as you might guess, simply tells us how well the left channel sticks to left-only programming (without "cross-talk" from the right channel) and vice-versa. Broadcast rules require only that the station itself maintain separation (at all audio frequencies from 30 to 15,000 Hz) of 30 dB. You don't need more separation than that (and probably a lot less) to get a good stereo effect.

The fact that a tuner brags about its 40 dB separation capability is a bit academic. Besides, the figure is usually quoted only for an audio frequency of 1000 Hz, where it's relatively easy to achieve good separation figures in a tuner design. Ask some manufacturer how well his tuner separation is maintained at 10,000 Hz or even 15,000 Hz and watch his face get red! Far better to tell us what the distortion is when tuned to stereo broadcasting or what the signal-to-noise ratio is under these listening conditions. Perhaps as standards change manufacturers will be required to state those facts as well.

Non-Performing Features

There are many useful features built into some tuners that don't particularly improve audible performance but are nice to have. I call them convenience features—and like all extra conveniences they cost money. There are also some features built into tuners that are absolutely useless; they usually cost money too. Single or twin tuning meters often show up on the face of a tuner. If one is supplied it usually indicates peak signal strength and is only slightly helpful in accurate station tuning because the "peak" reading hovers and isn't well defined. This kind of meter is useful when you're orienting an outdoor antenna to make sure it's pointed in the right direction.

The more accurate tuning meter is called a center-of-channel meter. Its pointer is at mid-scale. As you approach a station it wiggles off to one side, returning to exact center when you've correctly zeroed in. This type is easier to use and provides more accurate tuning. More sophisticated tuning indicators such as lights that blink when you're perfectly tuned are also available if you're willing to pay the price. None of these devices offers any better sound—just convenience.

The same is true of so-called muting circuits, which cut out interstation noise when you tune between stations. When activated they may also cut out reception of the very weakest stations you may want to hear. So if a tuner is equipped with the muting feature it should also have provisions for disabling that feature.

You can buy such extra goodies as electronic tuning, pre-set station selection—even remote control tuning if you're willing to pay the price. You can buy tuners with small oscilloscope tubes on them. The waveforms displayed help you tune correctly, let you look at the wiggles of the audio program you're listening to and reveal the multipath interference effects we talked about earlier.

Just about every modern tuner will have some form of indicator (usually a word lights up) to tell you when you're listening to a stereo broadcast. Some even have a switch position which allows

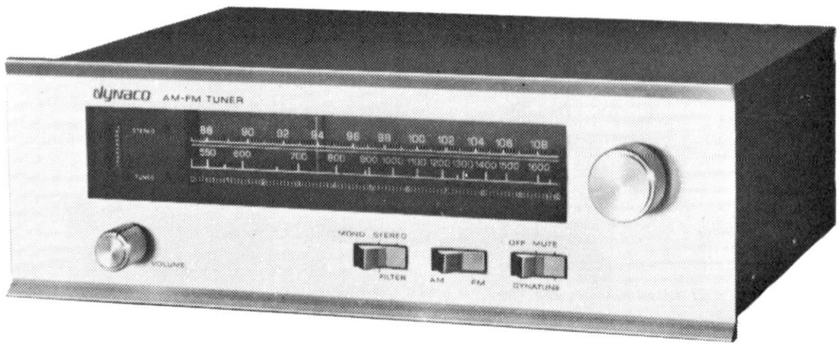

An AM/FM stereo tuner you can build yourself—from a kit.

you to tune to only stereo stations. The fact that the stereo light is on does not always mean you're getting stereo. Sometimes the dumb-dumb at the station forgets to turn off the "pilot signal" needed when stereo is broadcast long after he's switched to mono.

The newest gadget in FM tuners is called *digital readout.* Here, instead of a dial pointer and a bunch of numbers painted on a dial glass, the numbers pop up on computer-like readout tubes. *Very* impressive; also very *costly.* It contributes nothing to the sound you hear. I should qualify that. There is one form of digital readout circuit which is a direct result of the use of a new kind of tuning circuit called a *phase lock loop.* This kind of tuning (only a very few sets have it so far) is every bit as accurate as the tuning of the broadcast station itself. Since accurate tuning results in least distorted reception, this does improve sound indirectly. But digital readout as a mere substitute for a dial pointer and dial scale is hardly worth it and I'd look for better *performance* specs for that extra money.

As far as looking out for future advances, you might want to make sure that any tuner you buy today is equipped with a four-channel decoder output jack. This is a connection (usually at the back of the set) intended for a future four-channel broadcast system that *may* be authorized by the FCC some day. We'll mention more about this in Chapter 11, but just in case you're rushing out right now to buy your tuner or receiver, be forewarned. Also be forewarned that depending upon what kind of four-channel system the FCC approves the four-channel jack may or may not be useful.

Throwing Money Down the Drain

There's one small investment I urge you to make in order to realize full performance from whatever tuner or receiver you buy. I hate to call this item an accessory because in my view it should be a mandatory purchasing requirement for anyone getting into FM radio reception. A law ought to be passed requiring every tuner and receiver purchaser to invest in a reasonable FM antenna. This investment need represent only a small percentage of your total expenditure for electronics. You'll never regret it.

All the curves we mentioned earlier relating to sensitivity, signal-to-noise and distortion, realize their best values when a lot of signal reaches the antenna terminals of the tuner. But that signal must *reach* the tuner in the first place. In some experiments I ran using the piece of folded-up wire that most manufacturers toss into their tuner cartons—and laughingly call an antenna—I measured about 1/10th the signal strength that I was able to get with the same tuner connected to a decent outdoor antenna. In effect, my 1.8 microvolt sensitivity-hot tuner performed like any low-fi piece of junk with an 18-microvolt sensitivity rating when I used that piece of wire under the rug.

In a way manufacturers are partly at fault here. By giving you a poor excuse for a proper antenna they imply that this is all you will need for good FM reception. Nonsense! A decent antenna is one that's permanently and properly mounted outdoors as high as you can get it, and pointed in the direction that allows best, clearest reception. A decent antenna is fairly directional too, equipped with several elements like the one pictured in Fig. 5-7. The directional characteristics of the antenna help to reject those reflections that cause multipath, because the antenna is less responsive to signals coming from other directions than to signals coming from the direction in which the antenna is oriented.

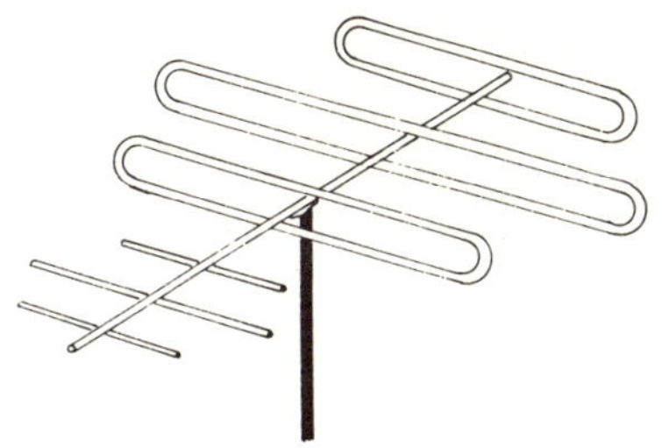

Fig. 5-7 A typical multi-element outdoor FM antenna provides better directional characteristics and signal-pulling power than simple indoor types.

Granted, some landlords don't permit installation of outdoor antennas. If you really can't sneak one up late at night, at least equip your tuner or receiver with a good indoor antenna—the popular "rabbit ears" type often used for TV—and orient it for best reception. If you have stations coming in from all directions, you might want to consider the use of a rotator, which can be controlled from a little box in your home and actually turns the entire antenna around and around until you get best reception.

Speaking of TV antennas it's all right to hook in to your existing TV antenna (a gadget called a *two-set coupler* is used to prevent interaction between the TV and the FM sets) providing your

TV antenna is not part of a professionally installed "master antenna system." Those installations often have "FM Trap" circuits deliberately included to reject FM interference from TV reception; you'd be worse off than if you had no antenna at all. While the length of the elements of a TV antenna is not quite right for FM, it's still much better than that piece of wire under the carpet.

How Good Is FM?

I've known hi-fi fans who have faulted the performance of their expensive tuners and receivers when actually the fault often lies with the FM station to which they're listening. The sad truth is that as FM has become more and more popular, FM station practices have become sloppier and sloppier. It's fair to say that the best tuners around today can reproduce signals better than most stations are transmitting them. There's not much your tuner can do if the DJ at the studio is playing a scratchy record with a bent phonograph stylus; or if his studio console isn't hooked up to a ground connection and he's broadcasting hum all over the place. If you're lucky and are within range of a couple of stations that try to do a good job (and there are still several around), great. If you're too far from such careful broadcasters, the solution, obviously, is to pull up stakes and move at once!

Chapter 6

AMPLIFIERS—THE HEART OF A HI-FI SYSTEM

As pointed out in Chapter 3, the amplification of audio signals (whether they be low-level phono signals, higher-level tuner or tape signals or whatever other program source you might have) can be performed by two pieces of equipment: the preamplifier and power amplifier. This job can also be done by a single piece of equipment known as an integrated amplifier. An integrated amplifier is really a preamplifier-control chassis and a power amplifier chassis all built together. Finally, the entire amplifier chain may be part of an integrated receiver, in which case the tuner will be built on the same unit as well.

The block diagram of Fig. 6-1 shows the function of each part of the preamplifier-amplifier component. Every one of the blocks shown is present in an amplifier, regardless of chassis format—single or multiple.

Preamplifier

Like so many other hi-fi terms, the word *preamplifier* has come to mean two things. Originally preamplifier meant that section in the amplification process that dealt with the very low-level signals associated with phonograph record playing, microphone am-

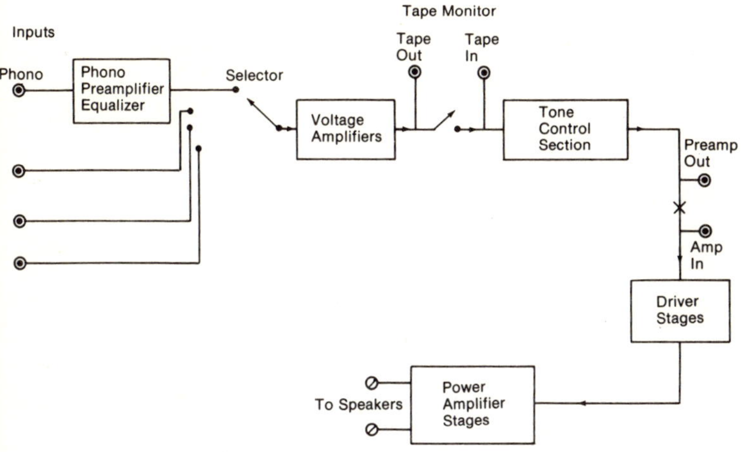

Fig. 6-1 Block diagram of integrated amplifier. Only one channel is shown, though the other stereo channel would be identical.

plification and amplification of the minute signals picked up by a tape-playback head on a tape recorder. The signal amplitudes from these three program sources are very minute: on the order of a couple of millivolts (thousandths of a volt). These signals therefore require *pre*-amplification to build them up to the signal level normally associated with tuner outputs or the outputs from complete tape-deck players (about 0.5 to 1.0 volt).

In the case of record playing, however, the preamplifier must do more than just amplify. When a record is cut it is deliberately recorded with frequency response that is anything but flat.

The graph of Fig. 6-2 shows that during the recording process low frequencies are purposely attenuated while high frequencies are actually *boosted* above the nominal zero dB reference level.

In order to elicit flat response from a magnetic phono pick-up or cartridge (the kind used in most hi-fi applications) a recording would have to have a groove with *constant velocity* undulations. That is, the needle would have to move at the same average *speed* at all frequencies. That would mean that low frequencies would have to be cut with greater lateral excursion in the groove than high frequencies because the needle moves back and forth fewer times per second and would therefore have to move further with each excursion in order to maintain constant velocity. Either the wiggles would have to be so great at low frequencies that one groove would run into the next one or there'd have to be far fewer grooves (and less playing time) on a record. To avoid this, the low frequencies are attenuated in amplitude during the recording

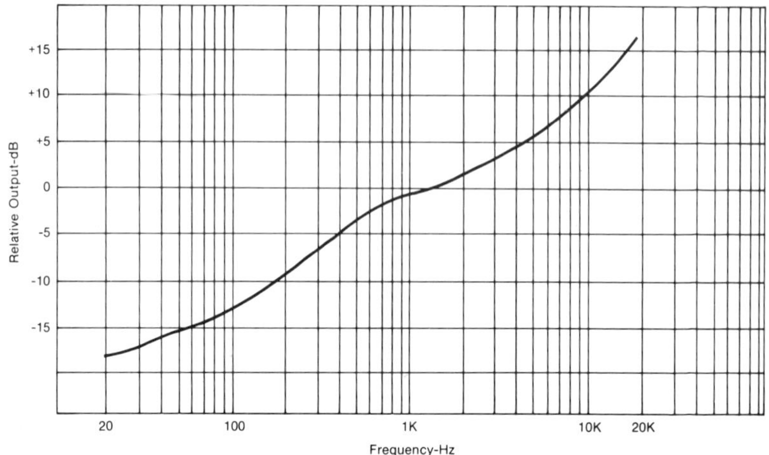

Fig. 6-2 RIAA recording characteristic.

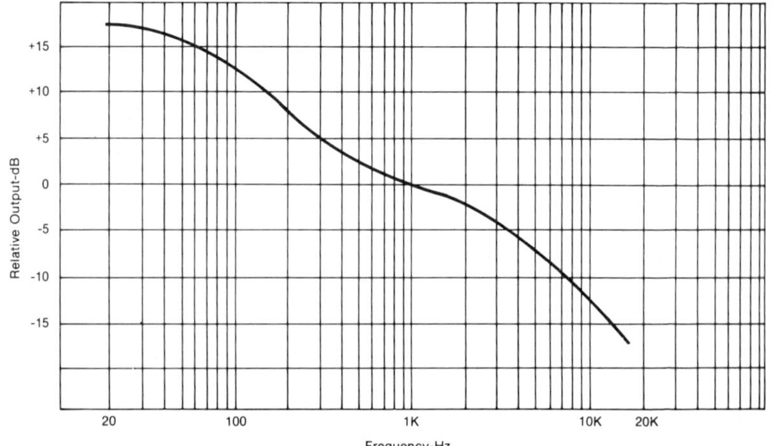

Fig. 6-3 RIAA playback characteristic.

process as shown in Fig. 6-2.

The high frequencies, conversely, are actually boosted during the recording process for reasons which will become clear when we talk about playback.

In order to end up with flat or uniform response the preamplifier, in addition to amplifying the minute signals picked up by the phono pick-up, must have a frequency response which is opposite to that used in the recording process, as shown in Fig. 6-3. If you add the "record" response to the "playback" response built into the preamplifier section, you end up with flat frequency response of the preamplified signal. These recording and playback response curves are called *RIAA* recording characteristics. The initials stand for Recording Industry Association of America. After years of non-standard recording techniques all record companies throughout the world now use this standard recording technique. As you look at spec sheets in choosing an amplifier you'll find that when it comes to phono, frequency response is given as "RIAA plus and minus 'so-many' dB" (e.g., Phono Response: RIAA ± 1 dB). This means that over the entire frequency range from 30 Hz to 15,000 Hz the response of the phono preamplifier will conform to the RIAA playback characteristic within the deviation (in dB) stated. Naturally, the lower the dB figure quoted, the better the spec.

Phono Hum and Noise

Because the phono-preamplifier section is dealing with such low-level signals, *hum* and noise is more of a problem than it is in tuner amplification, or the amplification of other "higher level" signals. Hum can be induced right into the phono cartridge (by playing the record player too close to such hum-inducing components as power transformers of another component, etc.) or it may be a function of the preamplifier design itself. Phono hum is usually quoted as so many dB below full output, and should be at least 60 dB if you don't want to be bothered by it. Since the hum is constant as far as the preamplifier is concerned, the signal-to-hum ratio will depend upon how much of a signal you get out of the cartridge or pickup in the first place. A pickup that puts out an average of 5 millivolts of signal will provide a better signal-to-hum ratio than one which only puts out 2 millivolts. It's important, therefore, that the manufacturer state his hum figure not only with respect to maximum power amplifier output but also with respect to phono-pickup *input* level. This is a great place for many manufacturers to "fudge" their results. For example, they may tell you that their product has a hum on phono of -70 dB, but not tell you that they're referring to a signal input of 10 millivolts (which hardly any pickup provides). If that hum figure were translated to a more typical reference input level of 3 millivolts, the hum figure would be only -60 dB.

High Frequency Equalization

Recall that in the record process we also boosted the high frequencies and, in playback, the RIAA curve calls for their corresponding attenuation. That's done to reduce record surface noise. Most of the bothersome surface noise is basically of a high frequency nature. Since treble content of music is generally lower in amplitude than middle or low frequencies, boosting their amplitude during recording does not push the grooves too far. The corresponding attenuation during playback, while restoring flat musical response, helps to lower the audible surface noise inherent in the record groove itself.

Phono Input Impedance and Overload

Most manufacturers will list two more characteristics of the phono preamplifier sections of their amplifiers. The input impedance for phono cartridges is usually given as 47,000 ohms or 50,000 ohms, which is just about right for most commercially available phono pickups. You should compare this number with a similar number supplied by the manufacturer of your phono cartridge, to make sure they're about the same. If the phono input impedance of the amplifier is higher than required, you may get a slight "peaky" sound in the extreme high frequency end of the spectrum; usually this is hardly audible.

The overload capability of the phono input is more important. While the typical phono cartridge may put out 2 or 3 millivolts under average program material inscribed on a record, during loud dynamic musical passages the signal picked up may be 20, 30 or even 50 times as great. The overload capability of a preamplifier tells how much signal the preamp stage can accept without creating distortion at that early point in the signal path. This kind of distortion has nothing to do with power amplifier output distortion, which we'll discuss later. Once the signal is distorted at the preamp stages up front there's nothing you can do in the amplifier proper (such as lowering the setting of the volume control) to get rid of it. So the higher the overload capability of the phono input the better. Values of 90, 100, even 200 millivolts are not unusual in good preamp sections.

Inputs, Switches and Controls

The broader definition of a preamplifier refers to the various switches, inputs and controls normally associated with hi-fi components.

The Rear Panel

On the rear of the unit you will find inputs for high-level signals usually called *auxiliary* (AUX) or *tuner.* You may find *tape input* jacks, *tape monitor* jacks and *tape out* jacks. One of the greatest sources of needless service calls on hi-fi equipment is misuse of the tape monitor switch and jacks. As you can see in the block diagram of Fig. 6-1, the switch labeled "tape monitor" is nothing more than a circuit interruption point. On either side of it are input and output jacks. This feature was originally developed for owners of good tape decks that have separate record and playback heads. One could connect from "tape out" to the input of the recorder for making recordings of any of the programs applied to the preamplifier. In addition, one could connect from the output of the tape deck to the "tape in" side of the tape monitor arrangement. In this way it is possible to monitor the resultant recording just a fraction of a second after the recording has been made. This gives more control over the finished recording than is permitted by merely listening to the signal going into the recorder. One can quickly make adjustments for over-recording by listening to the results.

Of course if you don't have a recorder connected in this manner and accidentally throw the tape monitor switch on, you have interrupted the signal path; nothing comes out of your loudspeakers. There usually follows a needless panic call to the service repair shop.

These days the tape monitor jacks and switch are being used for a lot of other things besides monitoring tape recordings. Since it is a convenient circuit interruption point, it makes a fine place to insert all sorts of accessory devices: special tonal equalizer accessories, four-channel adaptors and decoders or add-on reverberation units. Also, since this interruption point comes *before* the volume and tone controls, any recording you do by connecting your recorder to this point will not be affected by your tonal settings or listening volume, which can be adjusted to suit your taste while you are making the recording.

Some integrated amplifiers also have a means for separating the preamplifier-control section from the basic power amplifier section. This may be done by pulling out some wire jumper cables from back-panel jacks or by means of a little slide switch that sometimes comes on such products. This gives you much the same flexibility as owning a separate preamplifier-control unit and a separate basic power amplifier, even though both sections are contained in a single chassis.

A typical rear panel layout of an integrated amplifier is shown in Fig. 6-4. The rest of the items on that panel should be understood from the labelling of the diagram itself.

The Front Panel

A typical front panel layout for an integrated amplifier is shown

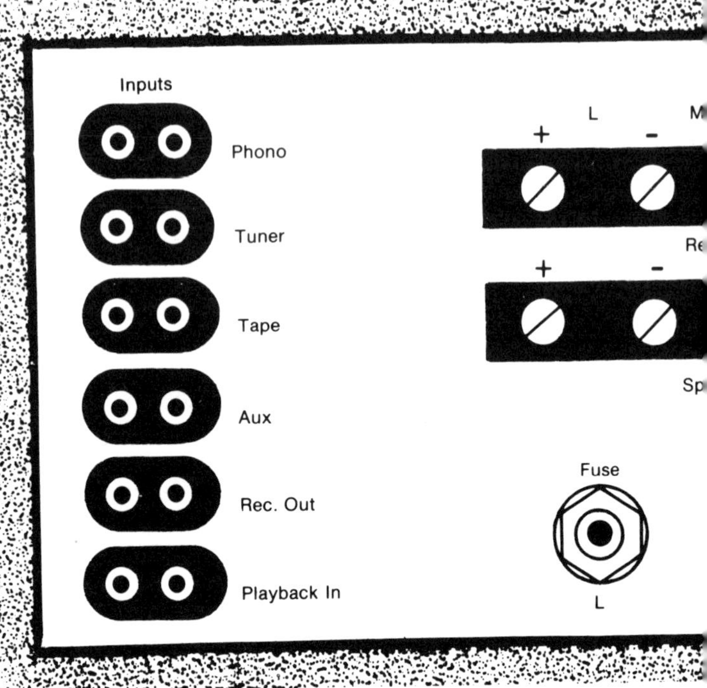

Fig. 6-4 Rear-panel layout of typical integrated amplifier.

in Fig. 6-5. The volume control simply controls the loudness of the signals applied to the power amplifier section; its use is obvious. The balance control helps you to set up equal volume from both speakers in a stereo set-up. The program selector chooses the program source and the power switch turns the whole thing on and off. Some amplifiers are turned on and off by rotation of one of the other controls, e.g., volume or tone. This is not as good as having a separate on-off switch which enables you to leave all your other controls at their optimized settings every time you turn the set on or off.

Tone controls

Many newcomers to hi-fi wonder about the need for tone controls. If indeed the end goal of all hi-fi equipment is to produce flat frequency response, why do manufacturers supply us with the means for upsetting that very flat response? There's really no contradiction here. The end goal *is* flat response but that means flat response all the way from the original signal to your ears. Not all speakers have flat response. Some

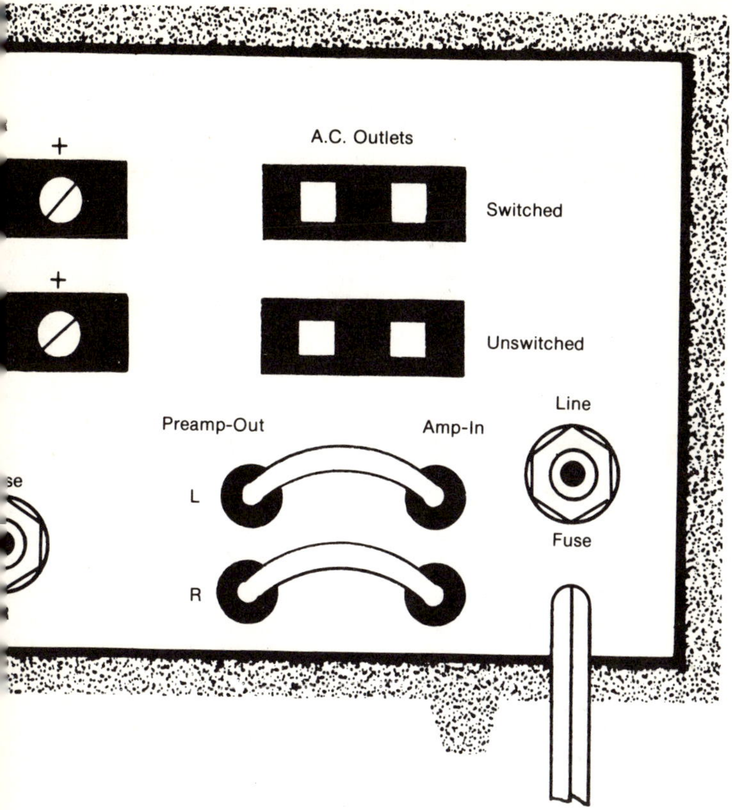

A.C. Outlets

+

Switched

+

Unswitched

Line

Preamp-Out

Amp-In

se

L

Fuse

R

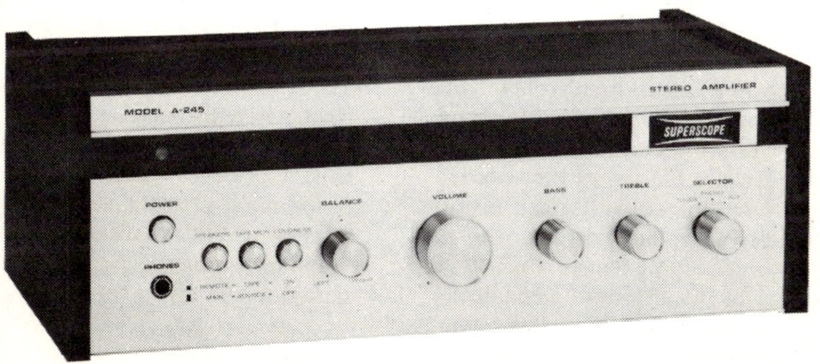

Fig. 6-5 Front-panel layout of typical integrated amplifier.

A high-powered basic amplifier.

rooms soak up high frequencies while others make highs sound overly bright. Some speakers actually produce too much bass when mounted on the floor instead of on a shelf; others do just the reverse. The best place to compensate for these deviations in response is at the control section of the amplifier—using tone controls. The trouble with most newcomers to hi-fi is that they tend to overuse tone controls. Show me a hi-fi user who has his bass boosted all the way full up (and maybe his treble control too) and I'll show you a first hi-fi set owner. Tone controls should be used in moderation to compensate for slight defects in the rest of the system. Used to excess they can easily drive amplifiers into distortion and produce unbalanced sound that is of the lowest-fi imaginable! Some manufacturers even supply a *defeat switch* for the tone controls with which you can knock them right out of the signal path.

The range of tone controls will usually be stated in ± dB at 100 Hz (for the bass control) and in ± dB at 10,000 Hz (for the treble). A typical figure is about 10 or 12 dB in each case. Alternatively, the manufacturer may publish a set of curves such as those shown in Fig. 6-6 or Fig. 6-7. The latter set of curves is preferred because with this tone control circuit you can alter the extremes of the audible frequency spectrum (by moderate use of the controls) without altering the middle frequencies much. Some tone control knobs operate for both left and right stereo channels simultaneously. More expensive units will usually provide separate control of left and right channels, usually by means of dual concentrically mounted knobs which can be turned separately or together.

Loudness-contour switch

When we listen to music at "lower-than-live" levels our hear-

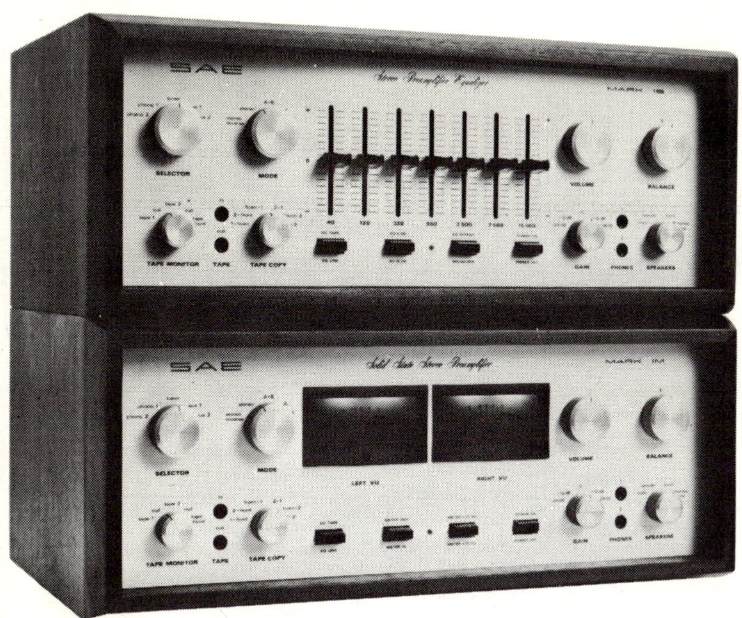

Separate preamplifier-control unit atop basic power amplifier.

ing tends to fall off faster at the extreme low and high frequencies than it does at middle frequencies. That's why background music levels often seem "thin" or deficient in bass quality. (It's also one of the reasons why many sound buffs tend to play their music so loudly.) To compensate for this effect, many amplifiers include a *loudness-contour* switch. With this switch in the "on" position frequency response will be varied depending upon the rotation of the main volume control. At high settings response will remain "flat" while at low settings extra bass and sometimes extra treble will be added, as shown in Fig. 6-8. This is a nice idea but one that doesn't always work perfectly. Sometimes you can hit live listening levels (where you don't need any compensation) even with your volume control set at its mid-point (depending upon your speakers, the program source, etc.). If you add loudness compensation under such conditions, the music gets too bassy and unrealistic.

Filters

High and low filters (sometimes designated as *scratch* and *rumble* filters) are also included in many amplifier control sections. *Rumble,* a low-frequency sound created by less-than-perfect turntables or record chang-

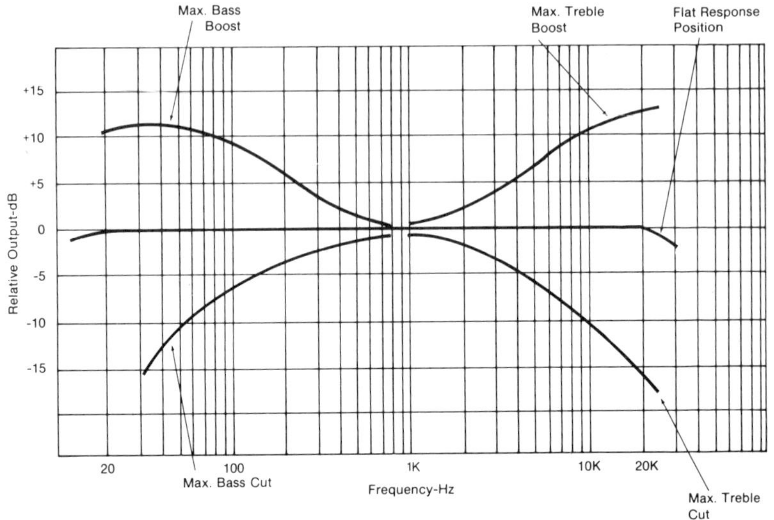

Fig. 6-6 Action of typical bass and treble controls.

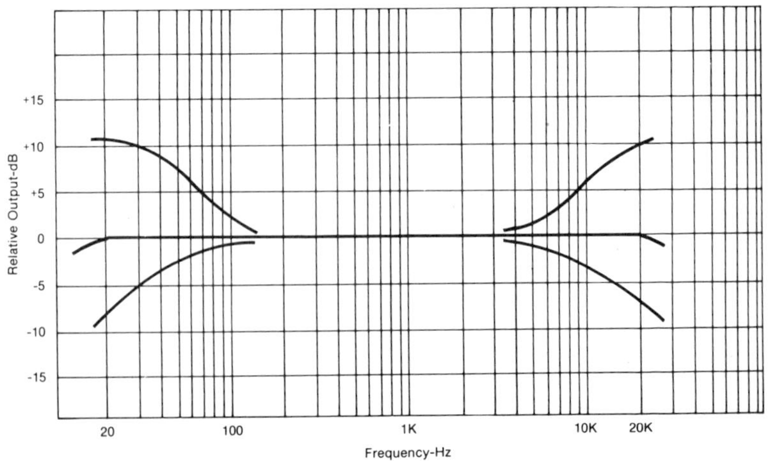

Fig. 6-7 Some tone controls permit adjustment of extreme audio frequencies without affecting mid-range tone.

ers, can be eliminated by these filters without seriously chopping out musical content. Record *scratch* (and sometimes FM background noise) can also be reduced by the use of that filter switch. The reason why these annoyances can't be properly eliminated by ordinary tone controls is that tone controls usually have a gradual slope in their action. Properly designed filters have a steeper slope, as shown in Fig. 6-9. They can therefore chop out the frequency extremes without affecting needed musical frequencies. Filters usually offer a cutting action of 6 dB or 12 dB per octave, from the frequency point at which they start their action. An octave is a doubling or halving of a frequency; thus 50 Hz is one octave lower than 100 Hz, and 10,000 Hz is an octave higher than 5000 Hz.

The other switches and controls shown in Fig. 6-5 should be understood from their names. Of course not every integrated amplifier or preamplifier-control panel will necessarily contain all of them (nor are all of them necessarily required in everyone's installation).

Amplifier Specifications

Among the many specs that an amplifier manufacturer will quote in his advertising literature are such things as frequency response, input level requirements for the different input jacks, hum and noise figures for each program source, power requirements, dimensions of the amplifier and the like. These are easily understood and don't require any elaboration. There are some terms, however, just as in the case of tuner specs, that need a lot of explaining. The most important of these is *amplifier power output.*

Amplifier Power

Electrical power is measured in watts; you'd think that it would be a simple matter for an amplifier manufacturer to tell you how many watts his product can deliver to a loudspeaker. Well, it used to be.

In the beginnings of hi-fi, engineers would simply feed a single tone (a simple sine-wave signal at one frequency) into an amplifier, keep increasing its amplitude and measure how many watts came out of the other end. All the while they would monitor the output to make sure it didn't get overly distorted. A reasonable distortion figure might be 1 percent. Most people begin to hear distortion when it exceeds that figure. In this case we're talking about harmonic distortion—the appearance of new signals which are harmonically related to the desired test tone. For example, if you overdrive an amplifier into distortion with a 1000 Hz signal it will start to give out harmonics of that frequency: 2000 Hz, 3000 Hz, 4000 Hz, 5000 Hz, etc. The amount of these unwanted harmonics compared to the amount of desired original signal is the *total harmonic*

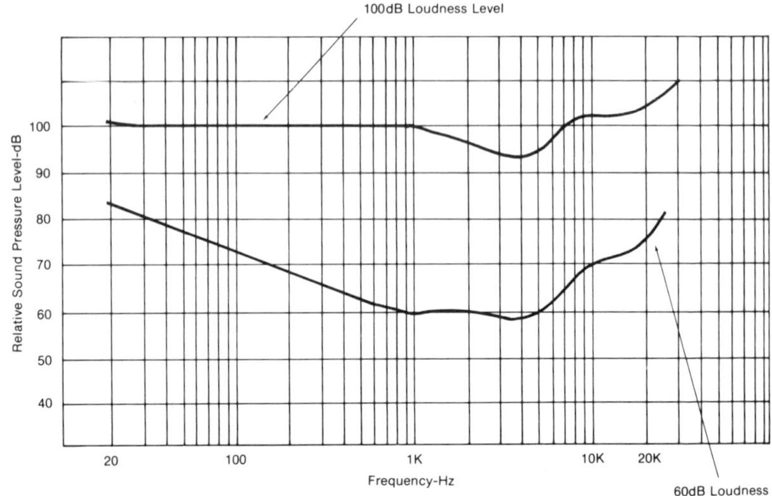

Fig. 6-8 Frequency response required at different listening levels to restore *apparently* **flat audible response.**

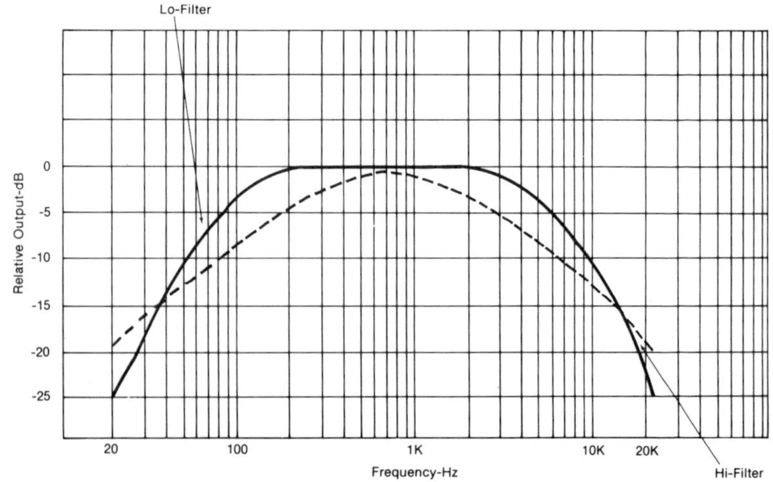

Fig. 6-9 Action of 12dB/octave filters (solid line) compared to action of conventional bass-cut and treble-cut extremes of tone controls.

distortion, expressed as a percentage.

Suppose in our example that the amplifier in question produces 10 watts per channel before reaching that 1 percent distortion figure. We'd rate that amplifier at 10 watts per channel *continuous power* (sometimes called *RMS power,* for root-mean-square) at less than 1 percent distortion. We'd also have to specify that the measurement was made with a test frequency of 1000 Hz. If we wanted to be really accurate we would have to specify that the speaker load into which the power was being fed had an impedance of 8 ohms (or 4 ohms, or whatever); power output capability will vary with speaker impedance. So far so good!

Unfortunately, music doesn't behave like a continuous sine-wave tone. It has moments of loud signals, moments of soft passages and so forth. Under these conditions, it was found that our Brand X amplifier could produce somewhat greater power per channel—say 12 watts—still with less than 1 percent harmonic distortion. This is true because the power supply in the amplifier (the circuits that supply the necessary operating voltage and current for the transistors and such) are not called upon to supply that current continuously. Under the more or less fluctuating demands of musical waveforms they stand up a little better.

It is this concept which gave rise to the term *music power* or *dynamic music power.* This rating is recognized by the hi-fi industry and its organization, the Institute of High Fidelity. Thus you may see this rating referred to as *IHF music power.* So long as it is accompanied by a rated distortion figure it is meaningful and legitimate. Beyond this point the power game gets tacky.

If you continue to increase the single test tone amplitude and feed it to our sample amplifier, the amplifier will deliver *more* power but the sine wave will begin to look different, as shown in Fig. 6-10. It's highly distorted; in fact such a waveform has about 10 percent harmonic distortion. If you ever read a power output spec without an accompanying

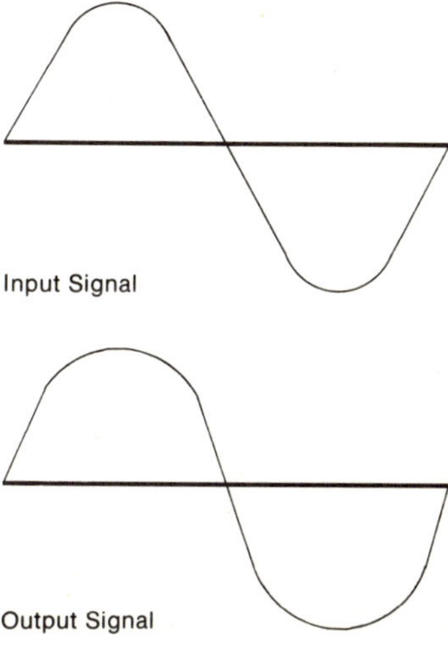

Fig. 6-10 Output waveform has about 10 percent distortion and no longer resembles input signal.

distortion figure, you can be pretty sure that the manufacturer has neglected to tell you that at that power figure distortion is about 10 percent or perhaps even more. Our friends in console radio-phonograph land usually figure power output on this basis; but they neglect to mention the distortion part of it. That way our amplifier is now able to produce 15 watts per channel!

If you carry this joke even further (which somebody did) and keep pumping in more signal to that poor amplifier, eventually you will reach a point where the output signal looks like the diagram of Fig. 6-11: an overdriven, clipped, totally distorted signal bearing scant resemblance to the one we started with. The wattmeter will read 20 watts under these conditions. The distortion meter (if anyone cared to use one) would read 50 percent—which means there is as much undesired signal coming out as there is desired fundamental signal! And that's how so-called "peak power" ratings were born. Totally meaningless but used by the low-fi people to promote inferior equipment purely on a high wattage basis.

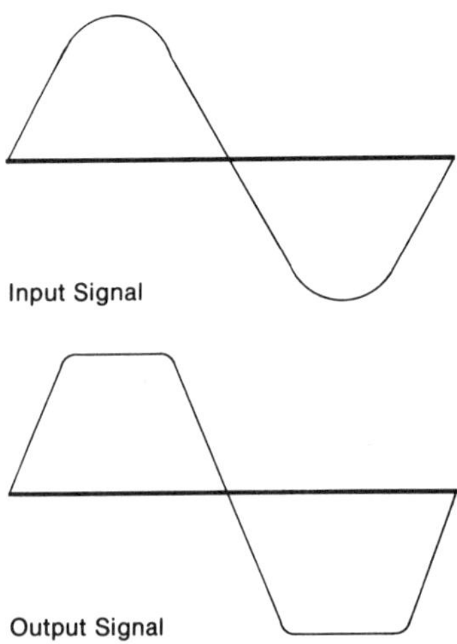

Fig. 6-11 Push an amplifier hard enough and it will produce more power output—at intolerable levels of distortion.

Now if you apply the idea of music power discussed earlier, with this idea of pushing for maximum power regardless of distortion, you come up with yet another term that has been seen in some ads: *instantaneous peak power,* sometimes abbreviated *IPP.* This could lead to a power rating of maybe 25 or even 30 watts per channel. If you double that (because we're dealing with a stereo amplifier) the final total power rating becomes 60 watts. Remember, we're talking about that same overworked 10 watt-per-channel continuous power amplifier we started out with. Nobody's changed a resistor or a transistor. Only the exaggerations have gotten bigger!

Stricter power ratings

The Federal Trade Commission has watched this circus for some time and is now finally planning to take some action, in line with the new consumer protection programs. They may limit statements of power to strictly continuous ratings, which the Institute of High Fidelity would favor (its manufacturers have never had anything to hide). Actually, a full specification would tell you even more than that. For a complete description of power output, you would need the continuous power capability stated not only for a single mid-band frequency but for all frequencies at which the amplifier is expected to work. For example, a really complete spec might read:

"Power output: 10 watts continuous power per channel, with 8-ohm speaker loads, at any frequency from 20 Hz to 20,000 Hz, with both channels driven simultaneously."

Accompanying that rating would have to be a statement such as:

"Total harmonic distortion: Less than 1 percent at any power level up to rated power output."

Power bandwidth

Since this kind of rating is not yet mandatory the IHF has another related rating which does tell you something about the power capability of an amplifier at its frequency extremes. That spec is called *power bandwidth.* It is simply the two end frequencies (highest and lowest) at which an amplifier will produce one-half its maximum *rated* power at its rated distortion. Since half-power means "3 dB less," it is sometimes stated as the frequencies at which power output is down 3 dB for rated distortion.

An example might be: Power bandwidth: 18 Hz to 40 kHz. This would mean that the amplifier (our same 10-watter) could produce 5 watts at 18 Hz and at 40,000 Hz with less than rated (1 percent in this case) distortion. Of course, if the full new specs for power rating become mandatory we won't need this power bandwidth spec any more because we'll know everything we need to know about power output capability over the entire audio frequency range anyway.

Power graphs

Sometimes power output will be shown in a graph such as that of Fig. 6-12. Here, at one glance, you can see what the distortion looks like at *any* power level. The solid line represents harmonic distortion—the kind we've been talking about. The dotted line shows another kind of distortion, called *intermodulation distortion.*

IM Distortion

Intermodulation distortion *(IM)* from an amplifier occurs when two simultaneous tones are fed to an amplifier at the same time. Under actual musical listening conditions this is going on all the time and the condition is therefore more representative of musical signals. When a high and low frequency signal are fed to the amplifier at once, the amplifier may produce *new* tones which are the sum and difference of the two starting tones. Thus, if you feed in 60 Hz and 4000 Hz you may expect an amplifier to produce a small amount of 4060 Hz and 3940 Hz and these two new undesired tones, expressed as a percentage of the total output, are called IM distortion. Most experts agree that this form of distortion is even more annoying than the harmonic kind. Obviously, the lower the IM distortion figure, the better the amplifier.

There is no direct relationship between how much harmonic distortion and how much IM distortion a given amplifier is likely to produce; but as you come close to the maximum power capability of the amplifier both kinds of distortion rise rapidly, as you see in Fig. 6-12.

Damping Factor

Another spec often noted by manufacturers is the *damping factor* of an amplifier. If you tap the cone of a loudspeaker with your finger, you'll notice that the sound emitted doesn't die down instantly. The speaker cone is, after all, a mechanical thing and continues to vibrate for some time after your finger leaves, depending upon how stiffly it's mounted and a lot of other factors. If the same speaker were connected to an amplifier and you tapped it again, you'd find that the "hangover" of

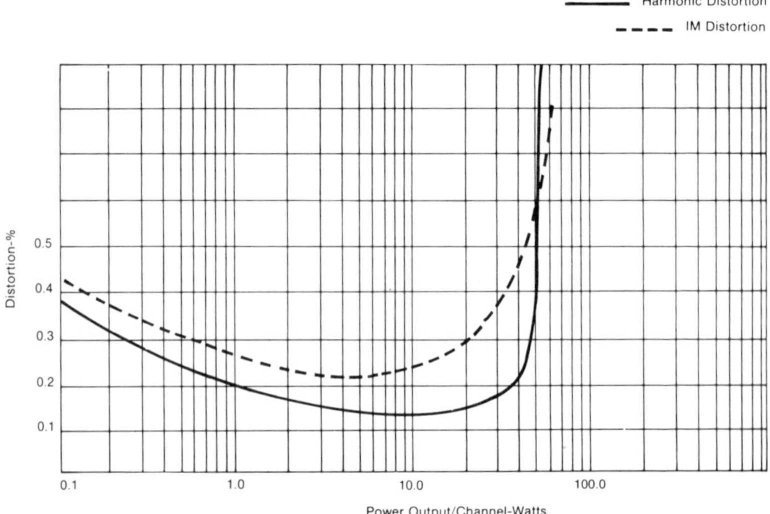

Fig. 6-12 Typical "power vs distortion" curve of an amplifier rated at 50 watts per channel at a rated THD of 0.5 percent.

sound would be reduced considerably.

The ability of an amplifier to "damp out" speaker overhang is called the damping factor and—within limits—the higher the number the better. However, once you get above 50 or so, any further increase in damping factor probably does not contribute audible improvement. Since damping factor is a ratio of amplifier internal impedance divided into speaker impedance it's important that the manufacturer specify at what speaker impedance the damping factor is being quoted. For example, an amplifier that boasts damping factors of 40 with a 16 ohm speaker will only measure 20 if referred to an 8 ohm speaker and 10 with a 4 ohm speaker. Furthermore, if you use very thin wire to connect your speakers to your amplifier and run them over great distances, the resistance of this extra wire becomes part of the internal amplifier resistance (as far as the speakers are concerned). The great damping factor built into the amplifier is in large part cancelled by the extra speaker-wire resistance. This is a good reason for using sufficiently heavy wire for long speaker runs. For distances up to about 10 feet or so 20 gauge wire is satisfactory. For longer runs 18 gauge wire should be used and for really long runs (to other rooms, etc.) even 16 gauge wire is a good idea.

Frequency Response

The *frequency response* of an amplifier or a preamplifier will be quoted something like this:

Frequency response:
20 Hz to 25,000 Hz ± 3 dB.

This means that over that range of frequencies, given a constant input amplitude, the output won't vary up or down by more than 3 dB at any frequency in that range. It's easy to see that if another amplifier lists:

Frequency response:
20 Hz to 25,000 Hz ±1 dB,

that's a better spec, since the variation over the stated range is less. The tough comparisons occur when you run into:

Frequency response:
25 Hz to 22,000 Hz ±1 dB

and are asked to judge that against

Frequency response:
18 Hz to 30,000 Hz ±3 dB.

In other words, the latter amplifier may very well be "down" only 1 dB at the end points (25 Hz and 22,000 Hz) of the first amplifier, or it may be down more or down less. There's no real way of knowing and there's no law that makes every amplifier maker use the same "end points" in defining frequency response.

There are really two schools of thought about frequency response of amplifiers (and audio equipment in general). One side believes that very wide response (well above and below limits of human hearing) is desirable, and they can give you all kinds of arguments to prove their point;

e.g., "I can't hear it myself, but my dog loves it!" The other side maintains that putting extra cost into an amplifier to get good response at those "unheard" frequencies is a wasted effort. I don't plan to settle this age-old argument here. Do some listening to both kinds and judge for yourself. Chances are that like me you won't be able to find any difference, providing all other specs are equal.

Other Things To Look For

Additional Speaker Connections and Power Receptacles

Many amplifiers have provision for connection of two or more sets of speakers so that the one amplifier can provide music in two or more locations. A nice feature —but remember that when you play two pairs of speakers at once, only half the total amplifier power is available for each location.

Amplifiers (and particularly integrated amplifiers) will usually have one or more AC receptacles on the back to which you can connect other equipment such as a tuner's line cord or that of your record changer or turntable. These may be labelled "switched" or "unswitched." The switched kind only have voltage at them when the amplifier is on. The unswitched kind have voltage available even when the amplifier is turned off. These are the ones to use for connecting your record changer. This ensures that if someone accidentally turns off the amplifier before the last record has finished, the changer will continue to rotate through its shut-off cycle and not create "dents" in the rubber idler wheel by being turned off while that idler wheel is pressing against the rim of the turntable. More about this when we get to record playing mechanisms in the following chapter.

Protection Circuits

When the "big switch" to transistorized or solid-state amplifiers was made some years ago, many early models often "blew up" without any warning. Frequently the problem was a short circuit accidentally placed across the speaker output terminals because the user had a frayed wire bridging across from one terminal to the next. The amplifier tried to deliver "infinite" current to an impedance of "no ohms" and blew its top. Since then all kinds of simple and sophisticated *protection circuits* have been developed to prevent such catastrophes. They range from simple fuses in the speaker output lines to very complex sensing electronic circuits which "turn off" the output transistors when excessive current or overheating is developed. The more complex the protection circuitry the higher the cost. Some form of circuit protection is a good idea and is usually less expensive than replacing costly output transistors (and other parts).

Power and Quality

While it doesn't necessarily follow that the very high-powered amplifiers should have better performance than lower-powered units, makers of the really high-powered jobs generally try to shoot for the very lowest possible distortion and the most features they can include to justify the higher prices of these units. Can you hear the difference between 0.5 percent distortion and 0.005 percent? Some experts say they can. Others argue that if you have much more power than you really need you're not likely to ever approach the power output point where noticeable distortion begins to occur. These arguments border on the philosophical rather than the technical. We'll leave all that to you and your auditioning of amplifiers.

Chapter 7

RECORD PLAYERS AND CARTRIDGES

Turntables

All a record turntable need do is rotate at a constant speed. That sounds pretty simple, but years and years of effort have gone into achieving that objective. We see new approaches to the problem appearing all the time.

Record players generally come in two forms: the single-play manual turntable/tone arm combination and the automatic turntable. *Automatic turntable* is a relatively new term for what used to be called a record changer, which over the years got to be sort of a dirty word. It implied poor quality and a mechanism that was inherently inferior to that of the single-play manual turntable. Today record changers are anything but inferior. In many ways they offer greater value than some manual turntables, but we'll get into that argument presently. Record changer manufacturers prefer to use the term "automatic turntable" to distinguish their product from its earlier inferior versions. That's the name that is beginning to stick.

Turntable Speeds and Rotation

Just about every turntable or automatic turntable will operate at at least two speeds: 33 1/3 and 45 rpm. All new records issued now use one of those two speeds. Popular "singles" are generally recorded in the 45 rpm format, while so-called "albums" or 12-inch LP records are designed to turn at 33 1/3 rpm. You will find some turntables that also include 78 rpm speed and even some that offer a 16 2/3 rpm speed. 78 rpm was the speed used years ago, before the development of the LP record by CBS Records. Since many people own these old records (some of them are collector's items), the 78 rpm speed persists on some equipment. The 16 2/3 rpm speed was originally proposed for very long-playing "talking book" records (for possible use by blind people), in which fidelity and frequency response would not be all that important. At such slow speeds it would be difficult if not impossible to get hi-fi wide frequency response; the slower the speed, the more times the needle or stylus has to wiggle back and forth in a linear inch of record groove passing beneath it. For some reason the 16 2/3 rpm idea never caught on and most manufacturers have eliminated that speed choice from their turntable and changer products in the last few years.

Wow

One kind of departure from true speed in a turntable is called *wow*. It is usually a variation that occurs once per revolution of the record and, because the pitch of the music changes at that slow rate, you hear a kind of wow-wow-wow sound which gives rise to the term. Wow is measured as a percentage variation from true speed. For a turntable to be considered good enough to be used with component hi-fi equipment that percentage ought to be no greater than about 0.1 percent.

Wow can be caused by an incorrectly manufactured disc as well as by the turntable itself. If a record is warped badly, the up and down movement of the tone arm and cartridge in trying to follow the record groove constitutes a cyclical change of overall speed that is heard as wow. If the grooves of a record are off-center with respect to the hole in the middle of the record, the tone arm will have to wobble back and forth, once per revolution, to follow the grooves. This too will be heard as wow. Before you blame your turntable or changer for the wow you hear make sure it's not the fault of the particular record you're playing.

Flutter

Flutter represents a change of speed too, but a much more rapid one as the term implies. Motors driving turntables generally operate at much higher speeds than the turntable rotates. Speed reduction is accomplished either by a series of stepdown pulleys or by a speed-reduction belt system. If a motor is rotating at 1800 rpm (typical of many phono motors) and this speed is reduced to 33 1/3 by one of the methods mentioned, variations of speed on a once-per-motor revolution basis may be transmitted as speed variation to the turntable itself. This gives rise to a flutter-like change of musical pitch that occurs at a much higher repetition rate. Again the term flutter is expressed as a percentage. In a good turntable or changer it should be less than 0.1 percent.

Rumble

Unlike wow and flutter, which both represent a change of speed of rotation (and therefore a change of musical pitch), *rumble* is the introduction of a new undesired sound as we listen to records. It is usually very low in frequency—sometimes lower than we can actually hear—but it can be so great in amplitude that it causes the loudspeaker to make large excursions back and forth and this can cause the audible musical or program frequencies to sound distorted. Sometimes you can't tell you've got a great deal of rumble until you play the same music on a system without it. The music suddenly sounds "cleaner" and less distorted. Rumble is caused by vibration (either of the motor, the turntable itself or a combination of the two) that is transmitted to the tone arm and pickup cartridge mechanically. It is then translated into a very low-frequency electrical signal that is then amplified by the electronics of your system and reproduced over the loudspeakers. Rumble is specified in two ways: *weighted* and *unweighted.* The weighted rumble number takes into account the fact that we don't hear extremely low frequencies as well as we do higher tones and tries to specify the aural effect of the rumble rather than its absolute numerical value. Naturally, the "weighted" spec will always result in a better number.

Rumble is stated in dB; the higher the number, the better. A good unweighted number would be 45 dB or better. That turns out to be about 55 dB in the case of the weighted specification.

Turntable Drive Systems

Indirect-drive turntables

One popular method of turntable drive (used primarily in manual single-play turntables) is the *belt drive.* As shown in Fig. 7-1, a rubber belt of proper dimensions is connected from the small shaft of the rotating drive motor to the outside of the turntable itself. Dimensions are so calculated that the high motor speed of rotation will be stepped

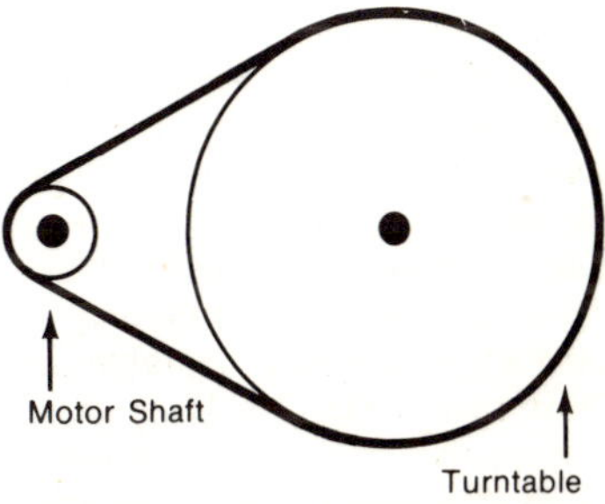

Fig. 7-1 A single-belt indirect-drive turntable system.

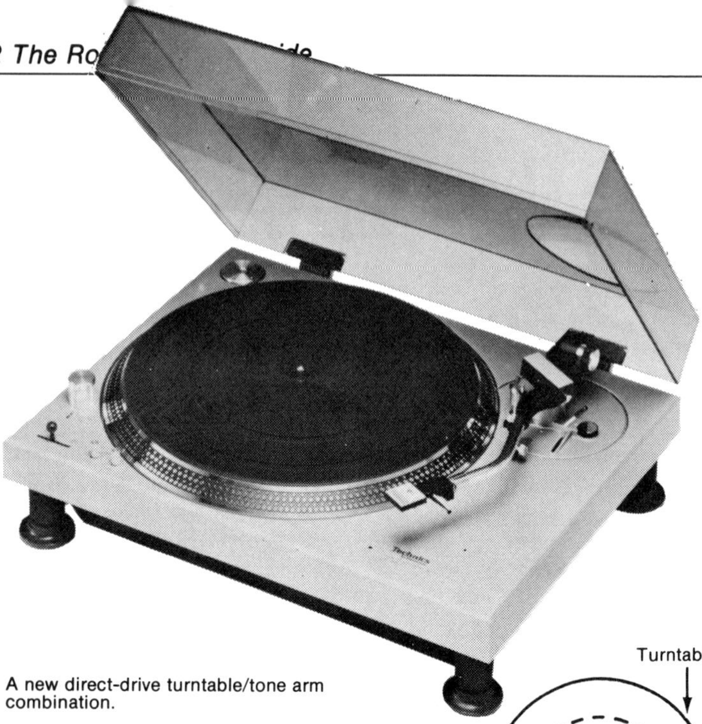

A new direct-drive turntable/tone arm combination.

down to the required 33 1/3 or 45 rpm speed. The use of rubber for this belt helps to isolate small speed variations (such as flutter) of the motor. Generally the use of a heavy turntable to give a flywheel effect also helps here. Some manufacturers build a turntable within a turntable (Fig. 7-2) to hide the belt.

This type of drive system is called *indirect drive* because the shaft of the motor does not come in contact with the turntable itself. Another form of indirect drive, and one most commonly used with automatic turntables, is illustrated in Fig. 7-3. Here an intermediate rubber wheel known as an idler wheel is caused to rotate by a motor shaft pressing against its edge. The idler wheel

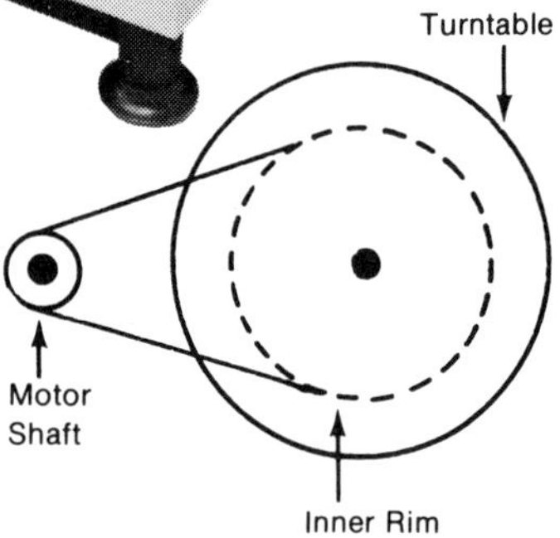

Fig. 7-2 The belt drives inner rim of turntable and is therefore not visible to user.

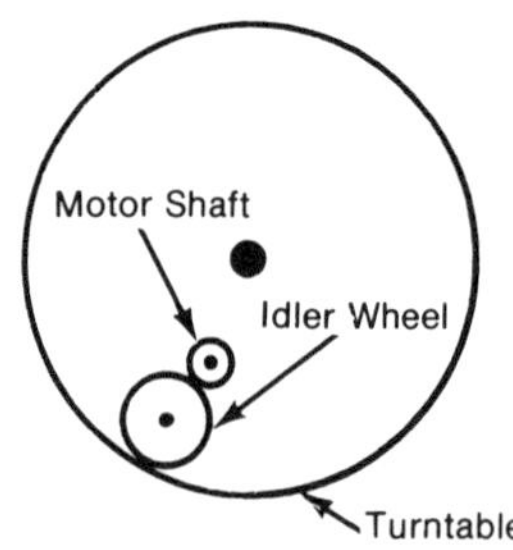

Fig. 7-3 One form of indirect drive uses idler wheel to couple torque from motor shaft to turntable inner rim.

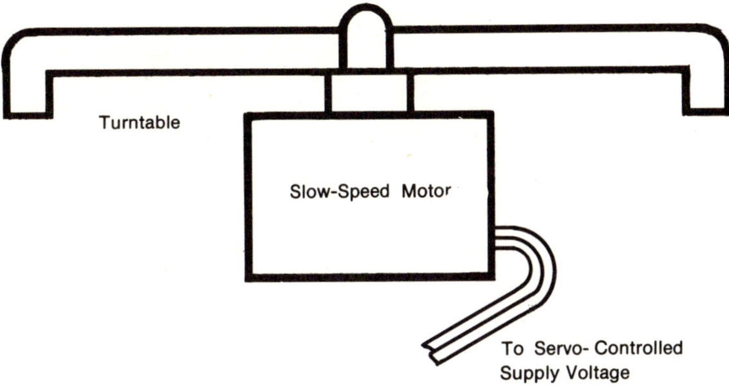

Fig. 7-4 The newest types of direct-drive systems use slow-speed electronically controlled motors.

in turn presses against the inside rim of the turntable itself. This system provides a simple way to change speeds because here the motor shaft has several "steps" machined to different diameters. To change speeds, the idler wheel is moved up or down to engage the shaft at different diameters. The smaller diameter of the shaft, when engaged by the idler wheel, will cause slower rotation of the turntable (say 33 1/3 rpm). Lowering the idler wheel to the largest diameter area of the motor shaft will result in a faster turntable rotation (78 rpm).

Direct-drive turntables

Recently new types of *direct-drive turntables* have become popular. These are high-priced models in the manual format only (so far) and involve the use of new, electronically controlled low-speed motors whose shafts are coupled directly to the turntable (Fig. 7-4). Accuracy of speed of these turntables is maintained by means of elaborate servo-feedback networks. Advantages claimed for these new models are extremely low wow and flutter as well as unvarying speed.

Turntable motors

Most record changers and manual turntables use one of two kinds of motors: *induction motors* and *synchronous hysteresis motors.* The induction motor is used mainly for lower priced turntables because its speed is likely to vary with load variations (number of records on the platter, for example) and with power line voltage. Some induction motors are equipped with sophisticated speed-regulating devices so it is not fair to say that all of them are inferior to the more expensive hysteresis types.

The speed of the synchronous hysteresis motor is locked to the incoming AC line frequency (60 Hz in the U.S.) and it is the type used in top-priced changers and turntables. (It's also the kind used in electric clocks, for obvious reasons.) With an induction motor, speed can be changed electrically by varying the voltage applied to the motor. This feature is sometimes incorporated in turntables and changers using the induction motor. If hysteresis synchronous motors are used, speed variation can be achieved only by altering the mechanical dimensions of shaft, idler wheel, etc. A tapered shaft could be used, for example, and the idler wheel could be manually moved up and down the shaft in order to make small speed adjustments.

Accuracy of speed can easily be determined by using a *strobe disc.* This is simply a paper or plastic disc onto which have been printed a series of radiating lines, closely spaced. When the disc is placed on a rotating turntable and illuminated by a fluorescent light, the lines, stroboscopically affected, appear to stand still. The light is actually blinking on and off 60 times per second. Some of the better turntables and changers incorporate this strobe light and calibration.

Manuals Versus Automatics

Probably the oldest debate in the hi-fi fraternity has to do with the merits and advantages of manual turntables as against the convenience and ease-of-use of automatic record changers. If I had written this book five or ten years ago I would have voted for the manual single-play turntable hands down. I'd have cited less wear and tear on the records, lower tracking force requirements, less wow and flutter and more rugged durable construction.

If you examine the specs of a modern automatic turntable and compare them with equivalent quality manual units, you'll find that all but the last of these arguments falls away. Better automatics have low wow and flutter today and they no longer require the heavy tracking forces they once did. Speeds are extremely accurate and often adjustable. Tone arms included with the better automatics have as little pivoting friction as some of the separately purchased tone arms used with single-play tables.

The choice is not even one of price per se. You might think that automatics, since they include all manner of levers, linkages and other parts needed for the changing cycle, are therefore more expensive—all other things being equal. The fact is that from a manufacturing point of view, all other things are *not* equal. Record changers, even the very good

A high-quality automatic turntable from Garrard. It can play one record at a time or serve as a record changer.

ones, outsell single-play turntables by far. Therefore they are produced in mass-production quantities and their cost tends to be lower. This fact offsets the lower cost of parts for a high quality manual turntable.

The durability factor—the one remaining advantage of manual turntables—is owed to the fact that there are fewer parts to go wrong in manual turntables. Therefore the choice should be based on other considerations such as your preferred listening habits.

If you want the convenience of being able to stack five or six records on a changer, you should select a good quality automatic turntable and stop agonizing over it. Don't worry about your records scraping against each other as they drop down from the changer's spindle. The outer rims of modern discs are actually built up slightly. When records lie upon each other their grooves no longer come in contact.

If you prefer the simplicity of a single-play turntable get a good one and make sure you get a tone arm (usually sold separately) suited to the turntable you've chosen. Often tone arms and turntables come from the same manufacturer and are therefore compatible. Otherwise your dealer can advise you.

Tone Arms

Whether tone arms come as part of an automatic turntable or as separate units there are a few basic things you should know about them. Most high-quality tone arms are made of a light metal to reduce weight and mass. Shapes can be grouped into three categories (Fig. 7-5).

In each arrangement the shell that holds the cartridge is at an angle to the arm axis itself. This angle is called the *offset angle* and its purpose is to minimize tracking error.

Tracking error is the angle that the cartridge axis deviates from tangency with the record groove at any given point. Since the tone arm is pivoted at a point far away from the cartridge, perfect tracking angle will occur at only one point on the record with these types of arms. The lower the tracking error (stated in degrees) at other points the better.

Several attempts have been made to eliminate tracking angle error. One of these is the *articulated arm* (Fig. 7-6). In this arrangement as the arm moves across the record, a second arm connected to the cartridge shell itself actually rotates the entire cartridge shell. Thus the axis of the cartridge is kept tangent to the record groove being played.

Since different cartridges have different weights and require different tracking forces the tone arm will usually have a movable balance or counterweight at its far end. Tracking force can then be adjusted by first balancing the arm horizontally and then applying more weight by moving the balance or counterweight towards the front of the tone arm. Usually some form of graduated scale is provided along the arm length. Otherwise a small scale specifically made for the purpose can be used to "weight" the tracking force by letting the stylus tip of the cartridge rest on the platform of this miniature scale.

Some of the other refinements you'll meet in examining tone arms are *cueing controls* that enable you to lift the tone arm and set it down without touching it and an *anti-skating* adjustment.

The purpose of the anti-skate adjustment is to counteract the tendency of the tone arm to slide inward as a record is being played. Forces tending to pull the

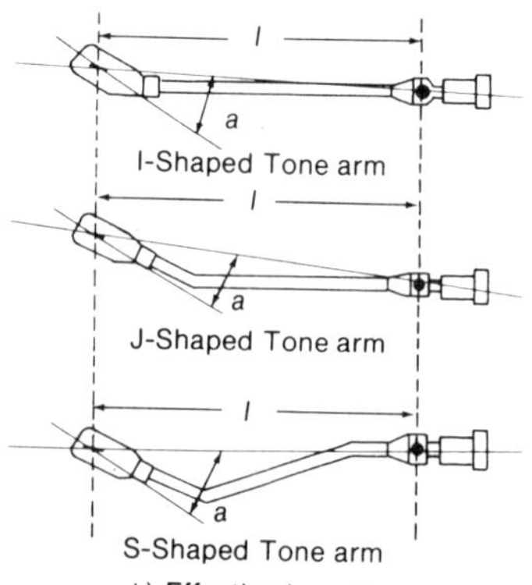

Fig. 7-5 The three most common shapes used in designing high-quality tone arms.

arm inward (arising from the geometry of the arm itself) would push the stylus against the inner wall of a groove and cause unbalanced record wear (or perhaps even audible distortion). Usually a small weight affixed to the rear of the arm applies a corresponding force in the other direction. This arrangement, called anti-skate, may even be made to correspond with tracking force adjustments.

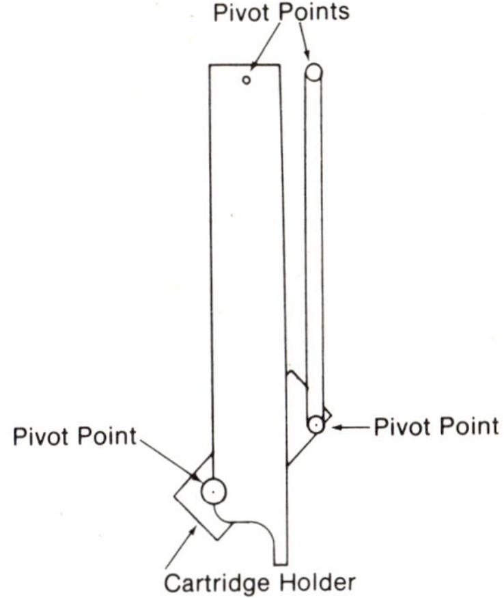

Fig. 7-6 The action of articulated tone arm keeps cartridge shell tangent to record grooves at any point in record.

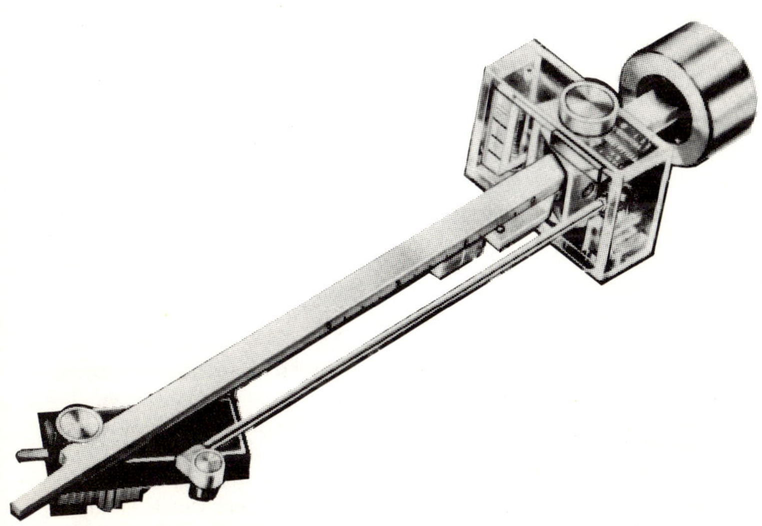

The Garrard Zero 100C articulated tone arm.

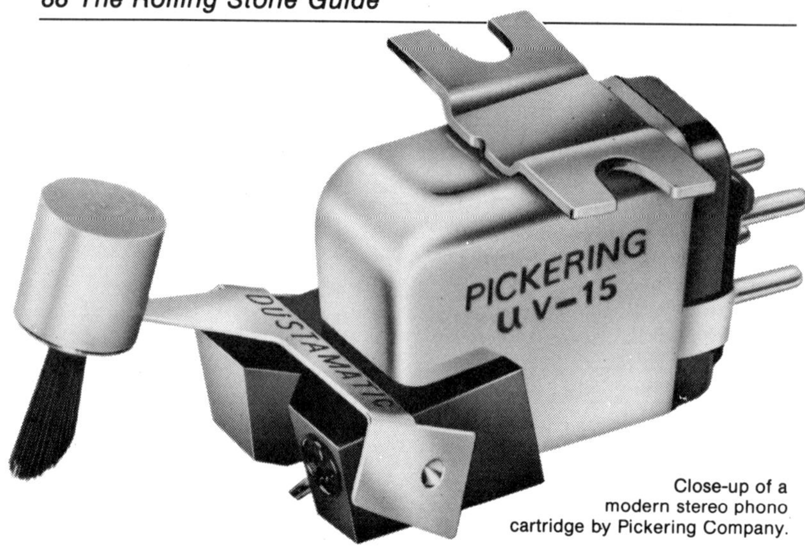

Close-up of a modern stereo phono cartridge by Pickering Company.

Cartridges

The *stylus* mounted in a cartridge at the tip of the tone arm has the job of changing into electrical signals the complex mechanical vibrations it picks up from the record groove. To do this the mass of the stylus and all associated moving parts has to be kept as low as possible; the greater the mass, the higher the inertia or resistance to quick changes in motion. An adequate downward force must be applied to the stylus to keep it in the record groove, but if this force is too great it will cause stylus and records to wear out quickly. The ability of the stylus assembly to move freely in both the lateral and vertical directions (stereo records involve both kinds of movement) is called *compliance* or *trackability*.

Earliest types of cartridges used a crystal material such as Rochelle salt. Such crystals, when subjected to pressure, produce a voltage or electrical signal. *Crystal cartridges* have generally poor frequency response, however, and very poor compliance. Therefore you'll find none used in high fidelity equipment. Children's phonographs still make use of them because they require less amplifying equipment and so reduce product cost.

Ceramic cartridges, especially modern ones, are a big step above crystal types in quality. Generally they are not as good as the magnetic types which are used almost exclusively in good component hi-fi systems.

Magnetic cartridges fall into three broad categories: moving magnet, moving coil or induced magnet. In the first kind a tiny moving magnet attached to the stylus assembly moves freely between stationary coils and induces a voltage in those coils (Fig. 7-7).

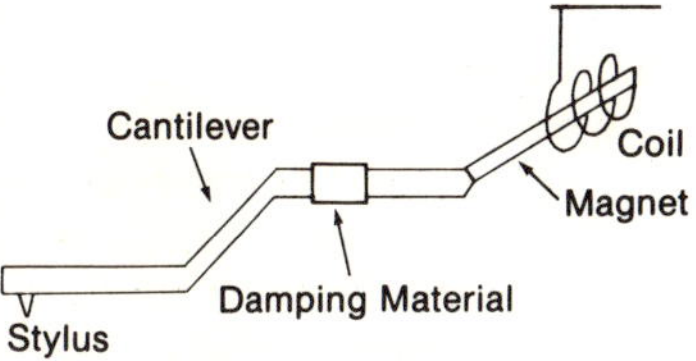

Fig. 7-7 Simplified view of stylus assembly used in magnetic cartridges.

The reverse idea is applied in moving coil cartridges. In them the magnet is fixed and the coils (one for each stereo channel) move in the magnetic field and have a voltage induced in them. The moving mass in this type of cartridge is usually lower than in the moving magnet variety but the signal output voltage is lower too and requires additional preamplification.

The induced magnet cartridges have fixed magnets and fixed coils (Fig. 7-8). Two small iron plates move in the magnetic field, causing variations in the magnetic forces and thereby in the voltage induced into the nearby coils.

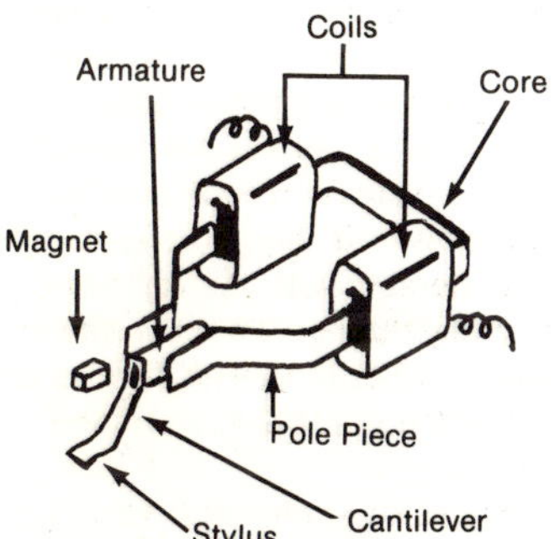

Fig. 7-8 Structure of *induced magnet* types of phono cartridges.

Normally, output voltages from magnetic cartridges will be anywhere from under 1 millivolt (1/1000th of a volt) to 7 or 8 millivolts when measured with certain standard test records having a known groove modulation. It is these very low signal levels that create the need for the phono preamplifiers discussed in the last chapter. A ceramic cartridge (and certainly a crystal one) would require neither a preamplifier nor equalization. On the other hand, the frequency response, compliance and other excellent performance characteristics associated with magnetic cartridges assure their universal popularity.

Cartridge Performance Specifications

Frequency response is the first specification you'll want to know about when selecting a cartridge. As with amplifiers you'll want wide response: at least from 20 Hz to 18,000 Hz or so, with no severe peaks or dips within that range. Any statement of frequency response for a cartridge must also contain a mention of deviation, expressed in dB; e.g., ±1 dB, ±3 dB, etc. The smaller the deviation in dB the flatter the response of the cartridge over its useful audio range.

Output voltage will usually be stated in millivolts (mV) and is important only in matching the cartridge to the amplifier. For instance, you wouldn't want to buy a cartridge that has 2 mV output and try to use it with an amplifier whose phono input sensitivity is 5 mV. To do so would mean you could not drive the amplifier to full power output with average program material. That would be like buying an amplifier of lower power output rating.

Tracking force is usually stated in a range of grams. It describes how much downward force must be applied to keep the cartridge from hopping around in the groove. If the tracking force stated is between 1 and 2 grams it would be a good idea to adjust your tone arm for about 1.5 grams. Any automatic turntable that requires more than 2.5 grams of tracking force doesn't deserve its name. On the other hand the long-term difference in record wear when tracking at 3/4 gram instead of at 1 gram is insignificant. If your record player is not well-isolated from floor vibrations (walking about, dancing, etc.) it is often advisable to track a bit on the high side of the recommended range rather than have the tone arm bounce all over the record as you move about the room.

Compliance is a measure of the stylus' ability to follow undulations in the record groove. It's expressed as the amount of deflection that a lateral (or vertical) force of 1 dyne applied to the stylus will cause. The larger the figure the better, although it also depends somewhat on tracking force. In a stereo cartridge, both vertical and lateral compliance must be stated. A good number is anything greater than about 10×10^{-6} centimeters per dyne.

Stereo separation describes the ability of the cartridge to separate the two program channels in a stereo record. It's expressed in dB, and will usually turn out to be greater than 20 or 25 dB if the cartridges are of good quality. Normally the manufacturer states separation at a frequency of 1000 Hz, where separation is easier to achieve. Don't expect to get that much at very high frequencies, where high separation is unimportant.

Stylus Replacement

Most hi-fi cartridges come equipped with diamond tip styli. While these diamonds are of industrial rather than jewel quality, they should remain in good shape for about a thousand record-playing hours or more. Only an experienced observer using a special microscope can tell you whether the tip of a stylus has become nicked or dull. You certainly can't determine its condition by running your finger across it. A worn stylus will of course increase record wear considerably and should be replaced —preferably with the cartridge manufacturer's own replacement stylus rather than with "universal equivalents" from questionable sources.

Stylus tips come in a variety of shapes. Conical or spherical ones are used with less expensive cartridges. Elliptical styli, with

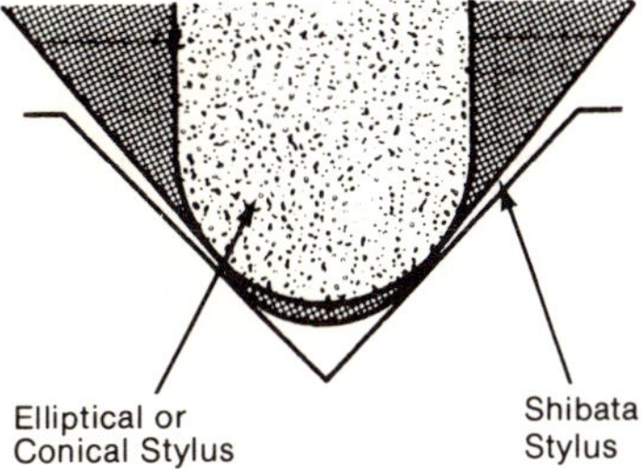

Fig. 7-9 The Shibata stylus contacts more of the groove wall as compared to a conical or elliptical stylus.

their greater dimension positioned across the record groove, usually deliver better response to higher frequencies. The smaller diameter of curvature perpendicular to the record groove enables the stylus to follow more closely spaced groove modulations. Since an elliptical stylus is more difficult to manufacture and install correctly it costs more than a conical type. Stylus force adjustment for an elliptical stylus is more critical than for a conical; you can do more damage sooner by using too much or too little tracking force with the elliptical tip.

The *Shibata* stylus (Fig. 7-9), a modified form of the elliptical, offers its advantages plus greater surface contact with the wall of the record groove (and thus less pressure per surface area). The Shibata, named for its inventor, was developed specifically for use in cartridges designed to play the new discrete four-channel records to be discussed in the final chapter of this book.

For the moment it's important to understand that while frequency response out to 20,000 Hz may be enough for the very best reproduction of stereo records, the newest breed of four-channel records (called CD-4 or Quadradisc) contains frequencies all the way up to 45,000 Hz. These high frequencies are not there for you to hear directly (even your dog would have trouble hearing at that frequency). Instead they figure in the decoding or demodulation process that results in the recovery of four separate and discrete channels from this new type of disc. If you aspire to a four-channel record playing capability in your system, it would be a good idea to purchase a cartridge that can properly track these new records. Such cartridges will also play regular stereo records perfectly—possibly with better fidelity than the best of today's stereo cartridges.

Chapter 8

THE MANY WAYS OF TAPE

Tape recorder captured from Germany during World War II. (Courtesy John T. Mullen)

Magnetic recording began in the 1920s when audio signals were used to magnetize a *wire* rather than the flat magnetic tape we use today. Recording on tape was highly perfected by the Germans during World War II. Shortly afterwards, an American named Alex M. Pontieff formed a company using his initials to form the name Ampex. One of the first major users of tape was Bing Crosby in the late 1940s and early 1950s, when he tired of doing his popular radio shows "live." Crosby became a heavy investor in Ampex, though he no longer has any interest in the company.

Before tape, recordings were made on stainless steel wire.

One of the first professional Ampex recorders, demonstrated by John T. Mullen. (Courtesy of John T. Mullen)

In addition to its use in audio recording, magnetic tape is responsible for all the prerecorded video programming you see and plays an important part in information storage in computers. The tape itself originally consisted of ferric oxide particles (known more popularly as *rust*) coated on a paper base. The relative weakness of paper as a base soon led to the use first of triacetate (or acetate) and later Mylar plastic (a trade name owned by Dupont).

Any tape recorder works along the same lines. The tape is moved at a steady speed across a tapehead, which is nothing more than an electromagnet whose magnetism is varied in accordance with the audio information to be recorded. The varying magnetic field is permanently imparted to the iron particles. When tape is played back, the magnetism previously recorded induces into a *playback head* a current which is then amplified to become the audio signal we hear from the loudspeaker. If you simply record an audio signal as we've described above you'll find that on playback it will sound severely distorted. Long ago it was discovered that if along with the audio signal you recorded a very high frequency one (at least five times as high as the highest audio frequency you wanted to record), the distortion was reduced, signal-to-noise ratio was improved and frequency response was extended. This extra super-audible frequency is called *bias,* and we'll get into it in more detail later on.

Open-Reel Tape Deck

There are three basic kinds of tape recorders now in use. The first in terms of age and quality is the *open-reel tape deck.* This machine uses 5-inch, 7-inch or even 10½-inch tape reels from which tape is threaded onto the machine much like movie film into a movie projector. On balance the open-reel machine has the best quality potential of the three types available. For one thing it operates at higher speeds than either *eight-track cartridges* or *cassettes.* That in itself means better frequency response. Many open-reel machines use three tape heads: one for erase, one for recording and a third for playback.

As the tape passes through it is first made blank by the "erase" head in case anything has been previously recorded on it. Then the "record" head applies the new audio information. If the machine is equipped with a separate playback head the resultant recording can be heard a fraction of a second later as the newly-recorded tape reaches the surface of that head. This is known as *tape monitoring.* If the machine is equipped with separate playback electronics you can actually compare the input program signal with the just-recorded tape by throwing a switch on the machine or on your amplifier or receiver (Fig. 8-1).

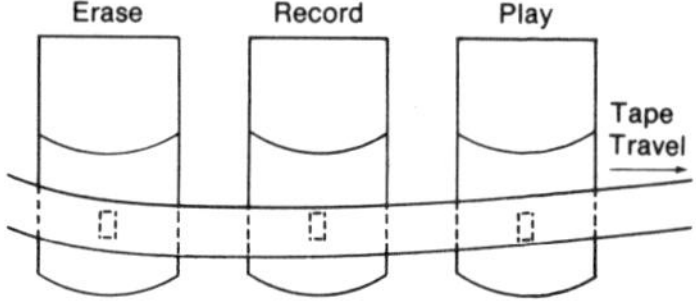

Fig. 8-1 Arrangement of tape heads in three-headed open-reel tape recorders.

The two-headed open-reel machine consists of an erase head and a combined record-playback head. Obviously with this type you lose the ability to do off-the-tape monitoring while recording; the single head cannot perform both functions at the same time. You'll find this limitation on most cassettes (which are really a kind of open-reel machine in miniature) and on some of the lower-priced open-reel machines.

In use, a playback head has specific requirements that differ from those of a record head. Thus the combination head is something of a compromise and can never provide quite as good recording or playback as separate heads.

The tape drive system on open-reel machines can be powered from a single motor (Fig. 8-2). There is a small-diameter wheel

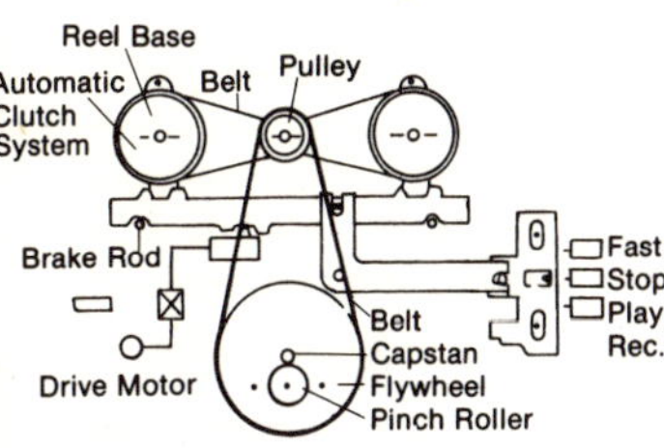

Fig. 8-2 A single motor can drive tape and both reels by means of belt and clutch arrangement.

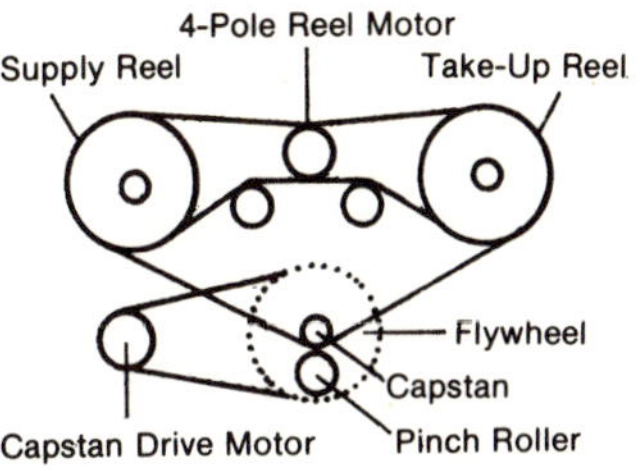

Fig. 8-3 In two-motor machines, one motor drives reels while second motor drives capstan-pinch roller combination (moving tape past tape heads).

and a large diameter wheel, between which the tape passes. These wheels are pinched together. The smaller wheel is the *capstan,* and is connected directly or indirectly to the motor; the larger wheel is appropriately called a *pinch roller.* The motor can be mounted almost anywhere since belts connect to the take-up and rewind reel hubs arranged in a rather complicated pulley system enabling the right belt to engage at the right time. Normally when tape is pulled from the outside of a full reel, it takes less pull than when it is coming from the inside diameter of an almost-empty reel. As a consequence a one-motor system, with its belts, tends to have more variation in speed between the beginning and the end of a tape. In carefully designed machines this variation can be held to under 1 percent—negligible, but there.

In two-motor machines (Fig. 8-3), take-up and rewind reels are moved by one motor while a second motor takes care of the tape drive itself. The most expensive open-reel machines (and even some of the newer cassette models) use three motors—one for each reel motion and a third for tape movement (Fig. 8-4). The quality of the drive arrangement, regardless of the number of motors, will determine some of the same performance specs which were of importance concerning record players, discussed in the previous chapter. Terms such as wow, flutter and speed accuracy apply to tape too, although there is no such thing as rumble or acoustic feedback.

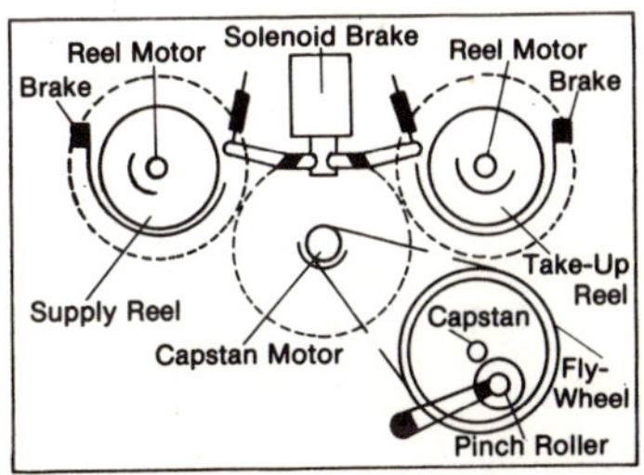

Fig. 8-4 In three-motor machines, each reel is driven by a separate motor in addition to the usual capstan motor.

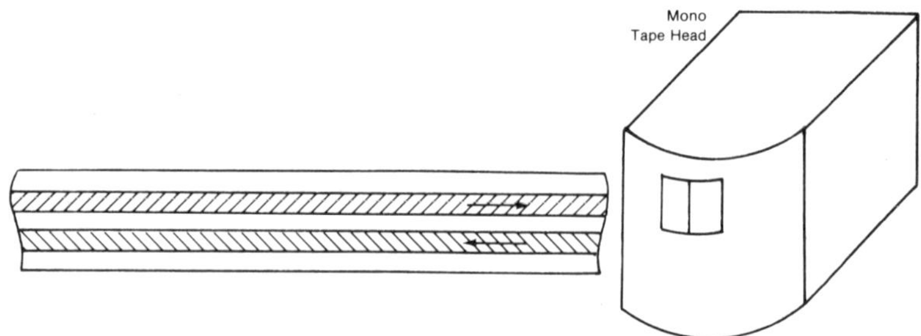

Fig. 8-5 Track and head arrangement used in mono tape recorders.

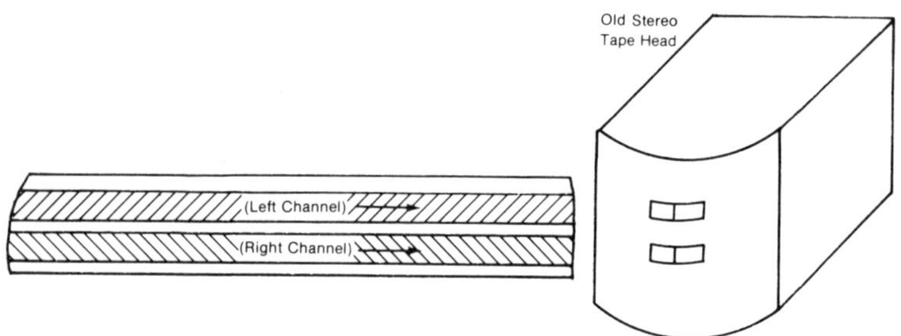

Fig. 8-6 In old stereo recorders and in some professional machines today, both tracks are used to record and play the two channels.

These are strictly mechanical problems and tape is entirely magnetic and electronic in its operation. Currently acceptable values of wow and flutter in open-reel machines of better quality would be under 0.1 percent.

Track Arrangements

The track layout of mono tape recorders of the open-reel type is shown in Fig. 8-5. The pickup tape head gap occupies half the tape width or a bit less. Thus a mono program can be recorded on the tape travelling in one direction; then when the tape is flipped over a second program can be recorded on the lower half (which is now the upper half), doubling the playing time.

Earliest stereo tape recorders had a record or playback head containing two separate circuits. Both halves of the tape width were used to record and play both channels in the same direction (Fig.

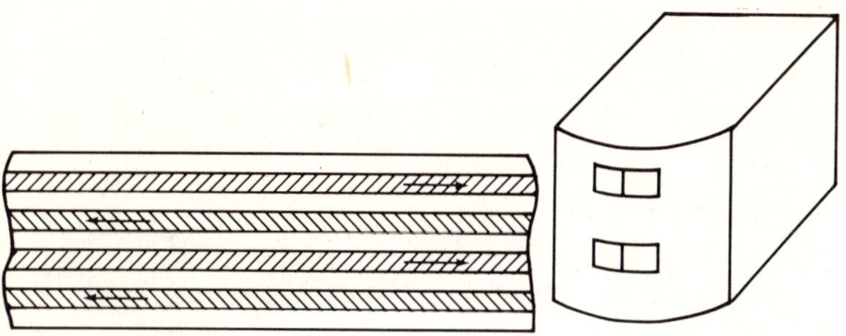

Fig. 8-7 Track arrangement used in four-track home open-reel stereo recorders.

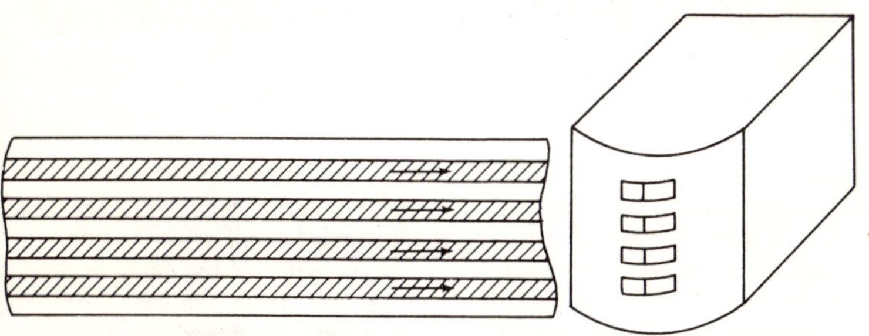

Fig. 8-8 This obvious track arrangement for quadraphonic cassettes would cut playing time in half and destroy compatibility between all cassettes. This arrangement *is* used in open-reel machines.

8-6). In time, head technology improved; pick-up "gaps" were made narrower and the four-track stereo tape recorder became standard. Please note that four-track is not to be confused with four-*channel.* The four-track stereo arrangement is shown in Fig. 8-7. With it the tape could be used in both directions once more to record or play two stereo programs—one in each direction—thereby increasing playing time to that available on mono machines. It is important to realize, however, that every time a track width is split in half, signal-to-noise capability is lost. This consideration stalks the present controversy over four-channel cassettes and explains why they aren't available as yet.

In 1970 when quadraphonic (four-channel) sound began to gain popularity, open-reel tape recorder manufacturers jumped right in and simply provided special new tape heads which had four separate gaps or magnetic circuits. All four tracks were recorded in the same direction (Fig. 8-8), again sacrificing playing time but in no way further degrading the signal-to-noise ratio capability of the system.

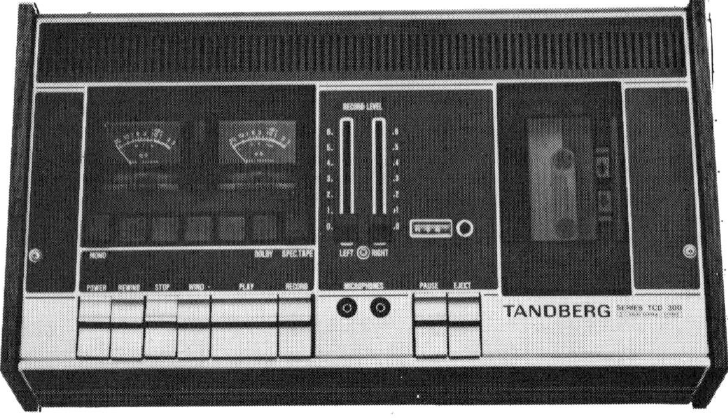

Cassette recorders, introduced for dictation, have come a long way.

Cassette Recorders

In the field of high fidelity, as elsewhere, things sometimes happen that you'd never bet on. When cassette recorders were introduced over a decade ago, no one would have predicted their popularity as a hi-fi component.

Consider the odds against them. They use tape barely ⅛-inch wide; yet, in stereo, four tracks are crowded into that width (two tracks in each direction, just like open-reel versions). They operate at only 1⅞ inches per second—the slowest of all the tape speeds. By all odds they should have absolutely the worst noise problem and poorest frequency response of all three tape formats. At the beginning they did, and hi-fi buffs resisted connecting these "toy dictating machines" to precious equipment.

Gradually technology solved the problem. Much work was done to make increasingly precise tape heads. If the gap on the head is made narrow enough, this helps to offset the lower speed. Some of the gaps on the new tape heads can barely be seen with the naked eye. Granules of ferric material made fine enough help improve frequency response and reduce residual tape hiss or noise.

These combined improvements created cassette units in the late 1960s that had signal-to-noise ratios of about 48 dB (just short of "true" hi-fi) and frequency response capabilities up to about 12,000 Hz (again just on the verge of true hi-fi).

The Dolby System

The single most important technological breakthrough was Dr. Ray Dolby's noise-reduction system. It changed the overall quality of cassette recordings and made cassette machines into a truly high-quality recording device. I'll explain how it basically works:

If you've ever attended a live concert, you may have noticed that when the very soft musical passages are being played, any audience snuffling, coughing, sneezing or even heavy breathing is clearly discernible. During loud musical passages, however, these extraneous soft sounds seem to disappear entirely. This is known as the *masking effect*. Our hearing mechanism has a sort of built-in automatic volume control that adjusts to the loudest sounds around us. (If it didn't, we'd probably all have long since been totally deaf.)

Dolby combined that knowledge with the fact that on tape it is the high-frequency noise to which we're most sensitive. Call it *tape hiss* for want of a better descriptive term. Dolby devised a recording circuit that changes the amplitude of the high frequencies to be recorded depending upon the level or volume of those high frequencies.

Fig. 8-9 shows that when loud musical passages are to be recorded all tones are recorded with flat response. As softer musical passages are encountered, the high frequency end of the spectrum is deliberately emphasized relative to the other frequencies. The softer the musical passage the greater the emphasis. You'll notice that at the bottom of Fig. 8-9 there is a cross-hatched pattern; this represents the "noise" that will be contributed by the tape and by the recording electronics of the tape machine.

Tape recording from discs. The units on lower shelf are Dolby noise-reduction components.

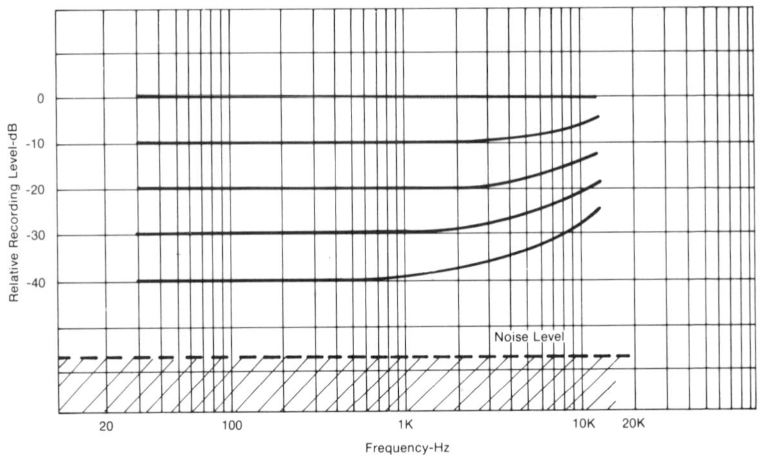

Fig. 8-9 The *encode* half of the Dolby system.

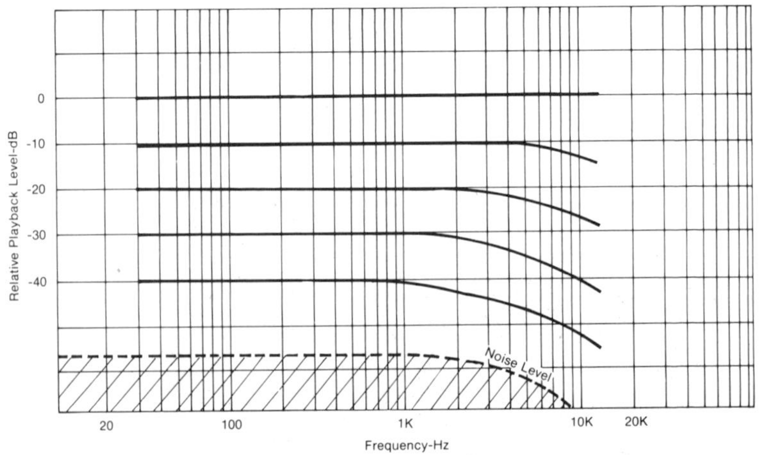

Fig. 8-10 The *decode* half of the Dolby system restores flat response while reducing tape noise.

Were you to play back such a *Dolbyized* tape on a regular cassette recorder the highs would sound somewhat shrill—especially during soft musical passages—because they've been deliberately boosted. If instead you play this tape back over a machine that incorporates a Dolby playback circuit, the action of that circuit (Fig. 8-10) will be just the reverse of that used during recording. Here, during loud passages, flat response is maintained. As soft passages are played, the circuit cuts or attenuates high frequencies by the same relative amount that they were boosted during the recording process. The result, musically, is flat response at any listening level for all audio frequencies. But look what happens to that noise pattern below: it drops right off the graph—just as if you had knocked down the highs with your treble control but without sacrificing musical frequency response. What an elegantly simple and effective idea!

The thing to remember about Dolby is that you can't "fix" a tape that's already been recorded without Dolbyizing. Once the noise is on the tape, it's part of the program. All Dolby can do is reduce noise that will be added to a tape in the course of making the recording. It can reduce it by about 10 dB in consumer versions of the Dolby noise-reduction system (known as Dolby B to distinguish it from a somewhat more complex Dolby A process used in professional recording equipment).

Furthermore, if you play a non-Dolbyized tape through a Dolby decoder unit, all you'll get is a program that's noticeably deficient in highs—particularly during soft musical passages. This 10 dB of signal-to-noise improvement pushed the noise right down from -48 dB to about -58 dB, surely a hi-fi spec. These days any stereo cassette recorder selling for about $200 or more is equipped with the Dolby feature.

Cassette Tracks

It was mentioned earlier that cassette tapes are divided into four tracks just like open-reel tapes. The tracks are laid out differently, however (Fig. 8-11): stereo pairs of tracks are next to each other this time. The first two are used for forward play and the bottom two are picked up by the tape head when the cassette is played on the flip side. This arrangement leads to a measure of compatibility that is not enjoyed by open-reel machines.

Philips Company of Holland (known here as Norelco), developers of the cassette format (and licensers to every cassette machine maker in the world), wanted the products to be totally compatible from mono to stereo and vice versa. With the track layout shown in Fig. 8-11 this is possible. If you play a stereo cassette on a mono machine, the mono head gap picks up the signals from both stereo tracks and you

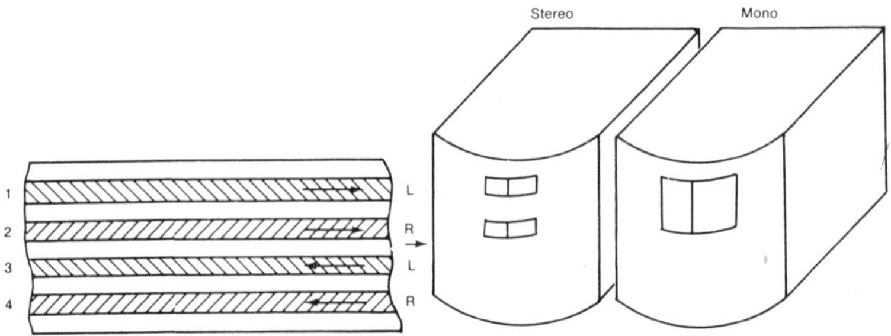

Fig. 8-11 Track and head arrangement for stereo cassettes.

get the sum of both signals in a mono equivalent. If a mono cassette is played on a stereo machine, the wider recorded upper track is picked up by both tape head gaps and circuits and you hear the same program from both speakers.

That's why any cassette recorded on any machine can be properly played on any other machine and you'll always hear the full program content in the right playing order.

Quadraphonic Cassettes

That brings us to the problem of recording four-channel sound on cassettes. The most obvious solution would be to record all four tracks in the same direction just as was done with open-reel machines. If this were done, however, and you played the tape back on a stereo machine, you'd hear only the front channels when playing the first side—thus missing the musical content of the back channels. Worse, were you to flip the cassette over to play the other side, you'd be listening only to the music of the back channels and you'd hear them playing backwards besides! When this track arrangement was first proposed by manufacturers, Philips quickly rejected it because it would destroy their much sought-after compatibility. They felt that it was the uniformity of cassettes that helped them reach their present popularity.

Instead, Philips proposed that the tracks be further split down to eight (Fig. 8-12). With this arrangment compatibility would be maintained all the way down to mono. Then the many licensees realized that each track would be but a few thousandths of an inch wide, requiring their tape heads to be and remain perfectly aligned if one track wasn't to cross-feed into the next. They refused.

Dr. Dolby, moreover, realized that further reduction of track width would again degrade signal-to-noise ratio. He said, in effect, that he wasn't going to bring cassettes two steps forward only to see quadraphonic tapes take

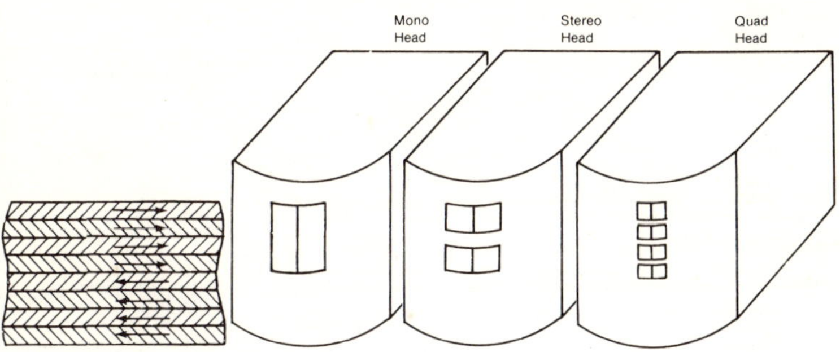

Fig. 8-12 This track arrangement for four-channel cassettes, proposed by Philips, would retain full mono-stereo-quad compatibility.

three steps backwards.

That's where things stood until very recently. As you might suspect, however, four-channel sound was too good a bet for the cassette people to miss. Many licensees have been developing new heads which would be fine and stable enough to satisfy the Philips proposal. One company has already announced a four-channel machine and others are sure to follow. At this writing, however, there are no pre-recorded four-channel cassettes available.

Eight-Track Cartridge Machines

The eight-track cartridge machine began life as a product intended primarily for use in cars. More recently it has found its way into homes as well—mostly for the purpose of playing cartridges that people bought to use in cars. The cartridge system was designed as a playback-only concept. There are very few home record/play machines. There are also very few companies making blank tape in cartridge form. Most dealers don't stock this kind of tape.

The pre-recorded eight-track cartridges you buy all use ¼-inch wide tape and, of course, eight recording tracks. The tape is almost twice as wide as that used on cassettes but there are twice as many tracks, each about as wide as a cassette track.

The track layout for stereo eight-track cartridges is different (Fig. 8-13). Here the tracks are numbered from 1 through 8; in the case of stereo cartridges, tracks 1 and 5 are used for the first set of musical numbers. The tape used in cartridges is an endless loop (Fig. 8-14), with tape pulled from the inside of the reel and returned to the outside circumference after it passes across the playback head. There are four stereo programs recorded, each using a pair of the eight available tracks.

A small bit of metallized tape passes across a switch at the end of the first program recorded on tracks 1 and 5. This tape ac-

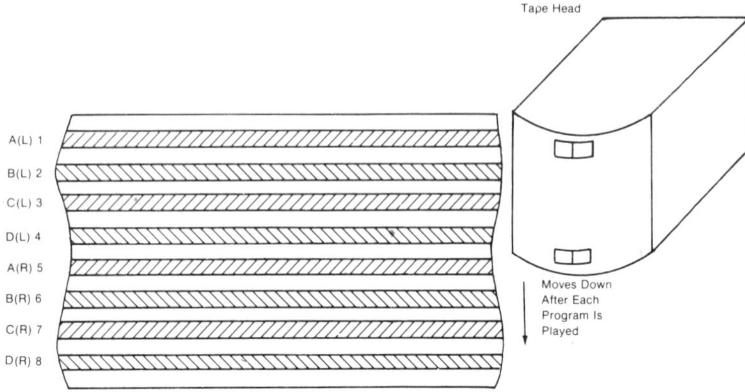

Fig. 8-13 Track arrangement used in stereo eight-track cartridges.

tivates a switch which activates a solenoid which in turn moves the entire tape playback head up (or down). Its action engages the next pair of tracks (2 and 6, then 3 and 7 and finally 4 and 8) until all programs have been played. Then the sequence begins all over again. Looking at Fig. 8-14 you can easily understand

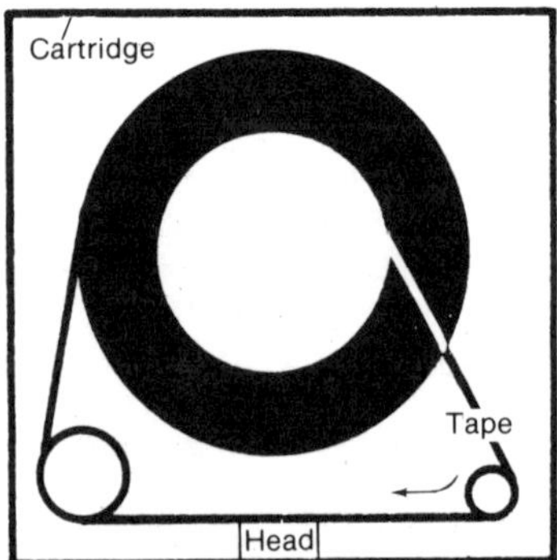

Fig. 8-14 Tape cartridges use an *endless loop* of tape, pulled from inside of the reel and returned to the outside.

that a rewind device on an eight-track cartridge machine is out of the question: it's impossible to stuff the tape back into the inside of the reel from where it's normally pulled. Some machines do have a fast-forward feature that enables you to reach a desired selection more quickly than if you had to wait for it at normal playing speed, but the fast-forward motion is slower than it would be for cassettes or open-reel tape. Most machines have an external button which advances the tape head to the next program tracks in sequence.

In terms of high-quality performance the eight-track medium ranks lowest of all three formats. Quality could be improved, however, since eight-track cartridge machines use wider tape and are speedier than cassettes. Perhaps the reason that not much emphasis has been placed on improv-

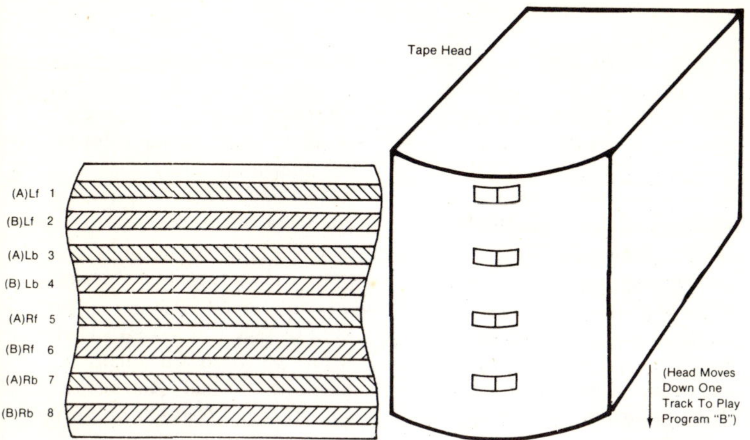

Fig. 8-15 Track and head arrangement used to play two programs (A and B) of Q-8 quadraphonic cartridges.

ing eight-track tape quality stems from the fact that it was originally (and still is) designed for cars. The noises of the road mask the tape hiss that would otherwise be so disturbing.

Q-8 Cartridges

The cartridge format does lend itself to quadraphonic adaptation; RCA first introduced what came to be known as Q-8 four-channel cartridges. The eight tracks remain but they now consist of two programs of four channels each (Fig. 8-15). New machines equipped with proper heads play these new four-channel tapes as well as older two-channel cartridges. Correct tape head stepping is accomplished automatically because the new Q-8 cartridge housings are equipped with an extra notch that "tells" the mechanism whether to "jump" one space or two when the tape loop comes 'round to the starting point. Older stereo eight-track playback units cannot compatibly handle the new Q-8 cartridges. You will only hear two of the four channels at any position of the playback head.

Types of Tapes

Tape comes in various thicknesses and therefore varying amounts of tape can be wound on a standard 7½-inch reel or onto the miniature reels contained in a cassette. The most popular lengths of tape supplied in open-reel form are 1200 feet, 1800 feet and 2400 feet. They will play 32, 48 or 64 minutes respectively if recorded in one direction only; or double that time if recorded in both directions.

The code number on a cassette tells you about available recording and playback time. Popular types are C-30, C-60 and C-90 and the numerals in each case refer to the number of minutes available when the tape is recorded or played in both directions (as is always true for cassettes). You can buy 3600-foot lengths of open-reel tape and 120-minute length cassettes but in order to cram that much tape onto a reel or into a cassette the tape has to be very thin. It often tears when used. Splicing a tape that has torn in an open reel is quite simple (especially if you equip yourself with an inexpensive splicing kit); trying to splice a torn tape in a cassette—though possible—can be quite tricky and frustrating.

The backing material on which the magnetic particles are coated may be acetate or polyester (the real name for Mylar, mentioned earlier). Acetate is subject to some deterioration with age and can become brittle. Polyester-backed tape does not change characteristics with age but is subject to stretching if pulled too hard. Of course the thinner the backing, the more susceptible the tape will be to stretching. That's another good reason for avoiding tape-length bargains and staying with more reasonable lengths per reel or cassette.

The latest refinement in tape-coating compounds has led to the development of *Chromium-dioxide tape,* usually abbreviated *CrO_2 tape,* its chemical symbol. This kind of tape offers still better signal-to-noise ratios and somewhat better frequency response—but the difference is slight. Furthermore CrO_2 tape requires quite different biasing from the recorder itself. If a recorder is not equipped with a switch for changing the record bias the use of CrO_2 tape will result in a poorer recording. Machines properly equipped to use this newer kind of tape will have a switch whose positions are usually denoted as "normal" (for ferrite coatings) and "CrO_2" for chromium dioxide tapes. Some even have a third position which is used with the newer low-noise ferric oxide tape.

Tape Machine Maintenance

While a tape machine has no "needle" that can wear out, there is one thing that should be done to insure peak performance with any tape machine. The heads must be cleaned, preferably after about every 30 hours of use (either recording or playback). The coating of the tape can be accumulated on the head after a time, preventing intimate contact between tape and tape head surface and clogging the carefully-machined gap in the tape head. The capstan and pinch roller may also gather oxide particles with time and they too should be cleaned periodically. Ordinary rubbing alcohol, applied with a Q-tip or a bit of cotton wound on the end of a toothpick, cleans heads, and there are also commercially available compounds sold specifically for this job.

Over long periods of time heads may become permanently magnetized; demagnetizers are available which can cure this trouble. You may be able to borrow one from your audio dealer since you will probably use it rarely. In addition to this regular care the manufacturer of the particular recorder may recommend other maintenance procedures—motor lubrication, etc. These suggestions should be followed if you want extended trouble-free service from your recorder.

Care of Tapes

As an amateur recordist, you're sure to over-record some of your tapes at one time or another. Besides the fact that such tapes will sound distorted, you may also find that when you try to re-use them the "erase" circuitry won't do a thorough job. You'll still hear a faint version of the previous recording in the background. If that happens, you may want to *bulk erase* your tape with a bulk eraser—a powerful electromagnet that receives its energy from the AC power line in your home. Don't ever use a bulk eraser if you only want to erase *part* of a tape; the bulk eraser does its job over the entire tape placed beneath it.

Usually when tape is rewound or wound in the fast-forward mode, the layers of tape tend to be more tightly wound and so each layer is physically closer to its two adjacent layers. Sometimes this can cause a condition known as *tape print-through:* some of the magnetic pattern of one layer is transferred to adjacent layers. When you play such tapes, you hear an echo (or perhaps a preview) of an adjacent layer's musical content. For this reason professional recordists often store their tapes in the "played" position without rewinding. The next time they want to play the tape they use fast rewind because the tightly wound layers will not remain in that condition long enough to cause print-through. For cassettes, the same precaution can be used; i.e., if you've played side A and don't plan to play side B, leave the cassette ready to play side B without fast-rewinding to the start of side A.

Tape will probably never replace the much more popular record disc as a home-entertainment program source. Yet a good tape recorder used in conjunction with the rest of your hi-fi equipment can be a very useful machine and does offer its own advantages as a sound source. If you want to do any live recording it's really the only way to go. Many tape enthusiasts buy (or borrow) new discs and promptly transcribe them to tape, for repeated listening, thereby preserving the noise-free surface of the disc. So long as you do this for your own pleasure, without planning to sell the tapes, it's perfectly legal—and no one is going to hit you for royalties.

Chapter 9

TIPS ON PUTTING TOGETHER A HI-FI SYSTEM

With so many program sources and equipment options available it's not surprising that many newcomers to hi-fi approach the problem of putting all the pieces together with some hesitation.

Once you get all the components home the interconnection of a hi-fi system is really a snap and takes only a few minutes. Just about any tuner works fine with any amplifier (if it's the two-piece electronic route you've chosen). The same is true of hi-fi phono cartridges, open-reel tape decks, cassette machines and eight-track cartridge decks.

There are a few rules and precautions that should be observed. While most of them are spelled out in the instruction booklets that come with the equipment, a quick summary of them may prevent frustration for the novice.

Cables and Wires

Most components come with the needed quantity of interconnecting audio cables. An audio cable is a special kind of two-conductor wire. Instead of the two wires running side-by-side, as in ordinary power cords, audio cable has an inner conductor surrounded by insulation which is in turn surrounded by a metallic outer conductor or *shielded conductor.* It completely surrounds the inner conductor either by being run spirally around the first layer of insulation or by being braided around it (Fig. 9-1).

Audio cables are usually supplied with pin plugs at each end (Fig 9-2). The tip of the plug is connected to the inner conductor while the outer metallic sleeve is connected to the surrounding shielded conductor.

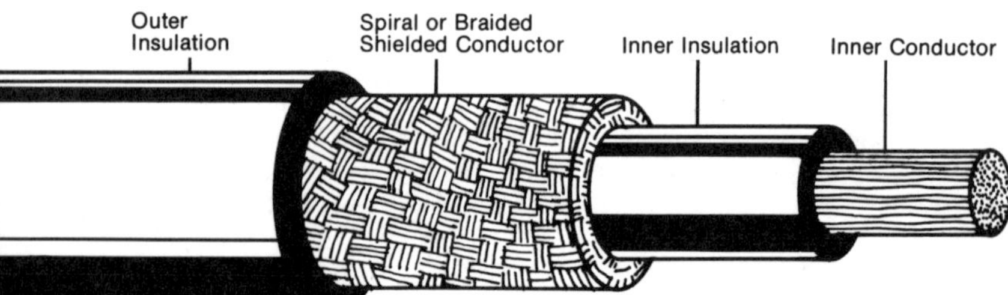

Fig. 9-1 The insides of an audio cable.

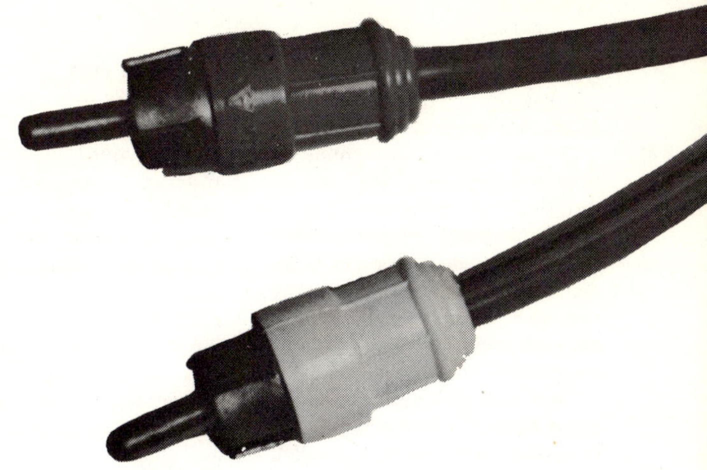

Fig. 9-2 Pin plugs are used for interconnecting hi-fi components.

Because the outer conductor usually is connected from the "ground" or chassis surface of one component to the other, the shielding arrangement reduces the hum signals which might otherwise be picked up from nearby appliances, fluorescent lights, etc. For this reason, audio cables are used whenever low-level program signals are to be connected from one component to another. Examples: connection of a phono cartridge to the phono inputs of an amplifier; of tuner outputs to amplifier inputs; of tape deck outputs to amplifier inputs.

When connecting low-level phono cartridge outputs from a record changer or turntable tone arm unit to an amplifier, the use of audio cables—however short they may be—is not enough to reduce hum to satisfactory levels. Remember that the signal produced from a magnetic cartridge is only a couple of millivolts. Even the introduction of a few microvolts of 60 Hz power line hum in the system will be amplified thousands of times to audible levels. Accordingly, an additional wire is usually connected between the metal base on which the record player mechanism is mounted and the metal surface of the amplifier or receiver chassis. If your system hums slightly when playing records, examine that connection. If it hums violently when you try to raise the volume level, chances are that one or both of your audio cables has an "open" ground connection. If you disconnect both cables from the amplifier and the hum stops, the trouble definitely lies in these cables or in the cartridge connection in the tone arm and not in the electronics of the system. Reconnect the two cables one at a time to determine which of the two might be at fault.

One good rule: *never* connect or disconnect cables while the equipment is turned on and the volume control is set above minimum. The sudden burst of hum that can be produced as you connect or disconnect a cable plug can pop a speaker voice coil or an amplifier output fuse.

Tape deck and tuner outputs are generally of higher signal level (from 0.5 volts to 1.0 volts) and therefore only regular audio cables are required to connect them to the rest of the system. Microphones come already supplied with shielded cables and plugs. Usually the microphones that come with tape recorders are high impedance types. Any attempt to lengthen the cable substantially will result in hum pickup.

If your amplifier is equipped with microphone inputs, you can connect your microphones to the amplifier instead of to the tape recorder mic inputs. Then, the high level inputs on the tape recorder are used to connect from tape recorder to tape outputs on the receiver or amplifier—the set-up of Fig. 9-3. In that way you can listen to the material being recorded over your loudspeakers or over a pair of stereo headphones. If your recorder is equipped with tape monitoring facilities, you can even monitor the results of your recording effort and prevent over-recording, distortion, etc.

Never try to make recordings by placing your microphone in front of your hi-fi loudspeaker systems. To record an FM program or transcribe a disc onto tape, feed cables from the tape-output jacks (sometimes labelled "rec out") of your electronics directly to the "line" of high-level inputs of your tape recorder. That way you avoid introducing distortion by the speakers and microphone as well as room sounds.

Speaker Connections

By the time program signals have been amplified by your power amplifier their levels are high enough to omit audio cables between amplifier and speaker terminals. Ordinary lamp cord suffices—the kind you can buy in any hardware store for a few cents a foot. Many audio shops sell lengths of packaged speaker cable which are supposed to be especially suited for speaker connection. The trouble is that in order to make the cable thin and flat, they employ very thin wire which has considerable resistance to electrical current flow. Its use for more than a few feet of run will cause dissipation in the wire of some expensive amplifier power when instead it might serve as useful electrical input to the loudspeaker system.

Even if you use lamp cord (sometimes called "zip cord") for very long runs, you ought to choose a higher-than-normal wire

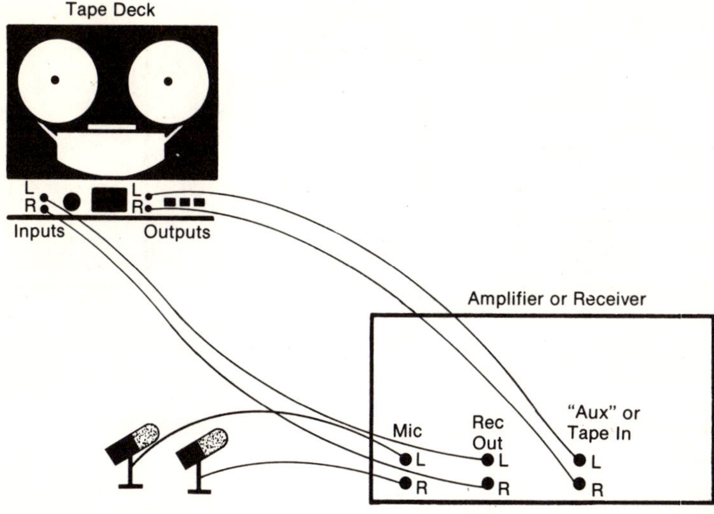

Fig. 9-3 If microphone inputs are available on amplifier, use these instead of inputs on tape recorder.

gauge. Specifically, for runs up to about 20 feet from amplifier to speaker, use No. 18 gauge wire. For longer runs use No. 16 gauge wire. If you plan to have speaker systems in more than one room, it is better to run wires from the amplifier to each speaker than to run wires from the first set of speakers to the second (Fig. 9-4).

If you can find the type of wire in which the two conductors are easily identified (either by color or by some coding on the outside insulation), this will simplify wiring both stereo loudspeaker systems "in phase." Thus both speaker cones will move in the same direction at the same time. Wiring speakers "out of phase" will result in reduced bass reproduction and confused directionality of musical instruments. Even if you think you have your speakers wired in phase it's a good idea to check this point by reversing the wires at the terminals of one of the two speakers. Listen to a monophonic program source while standing mid-way between the two speakers. If the sound does not seem to be coming from a point mid-way between the two speakers and if bass tones seem thinner than before, your first connection was the correct one.

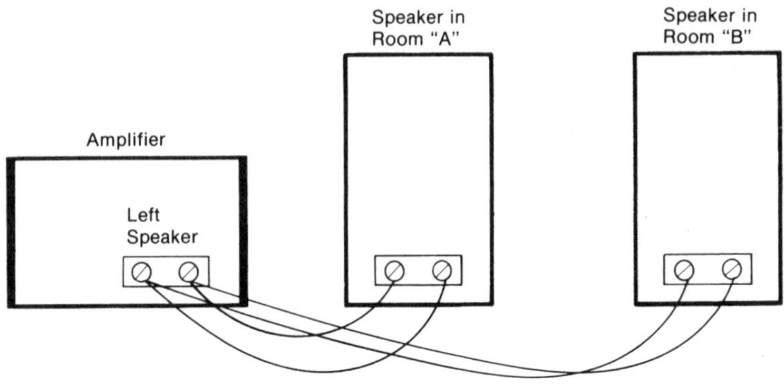

Fig. 9-4 It's better to wire multiple speakers (in different rooms) this way . . .

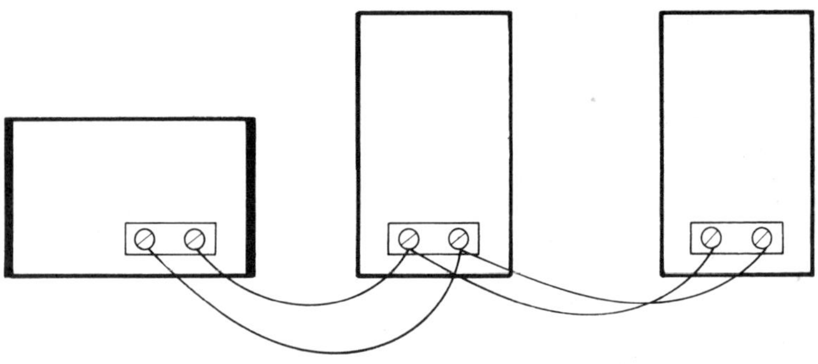

. . . instead of this way.

Where to Put the Components

Speaker placement in a hi-fi system involves some experimenting and considerations that we'll discuss in the next chapter. The electronic and other components of the system offer a great deal of placement flexibility but there are still certain precautions that should be observed.

Probably the fussiest component is your record changer or turntable. It should end up on a perfectly flat surface so that the tone arm exerts downward force only. There should be no side-to-side force to upset the delicate tracking and anti-skate adjustments provided on quality turntables and tone arms. Often the base on which you mount the automatic or manual turntable has provisions for adjusting the table so that it is level. If they are lacking or if you are mounting the turntable in a piece of furniture, use a carpenter's level to make sure this requirement is met.

Because the record-playing mechanism should be close to the receiver or amplifier (to avoid overly-long audio cable runs), all the electronics of your system should be as far from the loudspeakers as possible. If one of your stereo loudspeaker systems is positioned right near the record player, sound pressures created by that speaker when playing at loud levels can cause vibration of the tone arm and a vicious cycle begins that leads to *acoustic feedback,* a loud howl when you turn volume up to loud listening levels.

If you remove the cabinet from your receiver or amplifier you'll notice that at some location on the chassis there is a big black massive object—your power transformer. It is used to step down the 120 volts supplied to your home to the lower voltages required by the electronics of your hi-fi system. All that AC going through the power transformer means that there's a 60 Hz hum field in its vicinity. For this reason you'll want to position your record player tone arm as far away as possible from this side of the amplifier or tuner chassis. Also avoid too-close distance between tone arm and power transformer if you plan to mount the record player above or below the amplifier or receiver. Thus you'll prevent high hum levels.

Some tape recorders will operate properly whether mounted horizontally or vertically; others won't. Check the instructions that come with your recorder before you set it up permanently. The tape heads in your recorder are also susceptible to hum fields and this section of the recorder should be kept away from power transformers too. If you bought a separate tuner, preamplifier or basic amplifier, these too have power transformers. The same precautions apply.

As for record and tape storage: it shouldn't need emphasizing that records should be stored in their jackets and kept as clean as possible. Dust in record grooves is scraped along the groove by the travelling stylus and causes scratchiness and premature record wear. Several good record-cleaning devices are sold that both clean the record surface and apply an anti-static coating to prevent attraction of dust particles.

Recorded tapes also require careful handling (see "Care of Tapes," Chapter 8). Tapes properly stored in rooms of reasonable humidity and temperature should last a lifetime, provided you don't store them near strong magnetic fields (motors, fans, transformers, etc.) which might accidentally erase them.

Equipment Installation

The electronic components of your system are very versatile. They can be mounted in their own cabinet enclosures on an open shelf, in a book-case or on a table top, or installed in specially constructed furniture. Although today's solid-state equipment builds up far less heat than did the vacuum-tube equipment used years ago, some heat is still produced by power-output transistors, power transformers and other parts of the system. It's a good idea to leave plenty of ventilation around the components so that the heat can be drawn off by circulating air.

Never put one component above the ventilation slots or holes of another. This blocks air passage and can materially shorten the life of critical parts in the equipment. Most manufacturers will give you guidelines as to just how much space is needed around their equipment. If nothing is mentioned, leave at least six inches above any electronic unit free and a couple of inches at the sides and rear.

Powering Other Components

Usually your main amplifier or receiver will have one or more power receptacles on its rear panel. These are for connecting the power cords of associated equipment like tape decks and phonograph turntable motors. If the receptacles are labelled "switched" and "unswitched," connect the automatic turntable power cord to the one marked "unswitched." Then if you shut off the electronics, the automatic turntable will continue rotating until the last record has finished and the machine has cycled into its "neutral" off position. If an automatic turntable is stopped while playing a record, the idler wheel remains pressed up against the motor capstan. This pressure can cause permanent deformation of the idler wheel and increased wow and flutter.

Follow the Manufacturer's Advice

Hi-fi-component makers spend a great deal of time and money writing and publishing the very complete owners' manuals they include with their equipment. Letters that manufacturers get from customers make it obvious that many people do not bother with instructions. That's a mistake. The diagrams and clear language of these booklets can save you grief later.

After a first reading go back to the beginning and actually start the hooking-up as you read. Often the last page or two will describe what few maintenance procedures might be required. They will also list some of the common troubles you're likely to encounter and tell how to correct them. For the new (or even the old) hi-fi buff there is nothing so thrilling as throwing that power switch for the first time and having everything work.

Chapter 10

YOUR ROOM AS PART OF YOUR SOUND SYSTEM

In Chapter 3, I wrote that "all you really hear are your loudspeakers." Strictly speaking that's not entirely true. You also "hear" the room in which your sound system is finally installed. Is it a small room? Is it auditorium size? How is it furnished and what is its shape? All of these things have something to do with how your sound system will perform.

When large concert halls are designed they are built for their end use, the presentation of music. Despite increasing knowledge of acoustics and hall design, occasional blunders are made. When Philharmonic Hall in New York's Lincoln Center was finished over a decade ago, conductor Leonard Bernstein cringed noticeably when interviewed about the sound quality of the new hall. There were dead spots (places where you just couldn't hear the music properly) and other sonic defects immediately apparent to audience and professionals alike. Despite the fact that all kinds of money had been spent to make the hall acoustically perfect, much money was later spent "fixing" the hall so that now it isn't too bad. The Metropolitan Opera House, on the other hand, was designed according to successful classical European models and turned out perfectly.

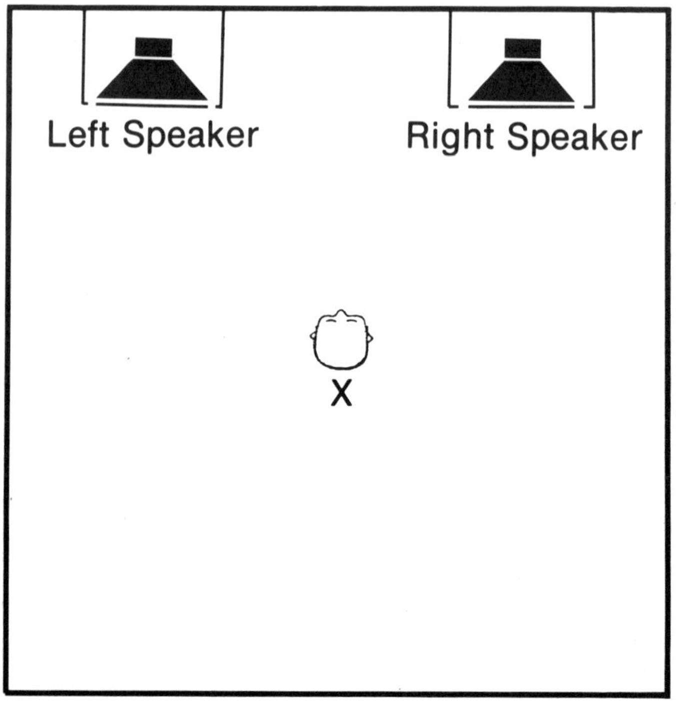

Fig. 10-1 The *imprisoned* stereo listener.

Your listening room, however, was not designed specifically for music reproduction. Few of us are lucky enough to have a music room built with that end purpose in mind.

I am always amused when I run across diagrams in hi-fi equipment manuals that show you how to set up a pair of speakers and tell you where to sit in relation to them. They usually look somewhat like Fig. 10-1. The room has no doors or windows and I've always wondered how that bald-headed listener ever gets in or out. The room also has only one chair—perfectly positioned at the X-marks-the-spot point shown in the diagram. The implication is that if you move away from that spot, your stereo illusion is destroyed. Nothing could be more untrue. Stereo sound is not that demanding.

The Long and Short of It

If you've got a square room it's not too important which wall you use to back your speakers up against. Merely avoid sitting right

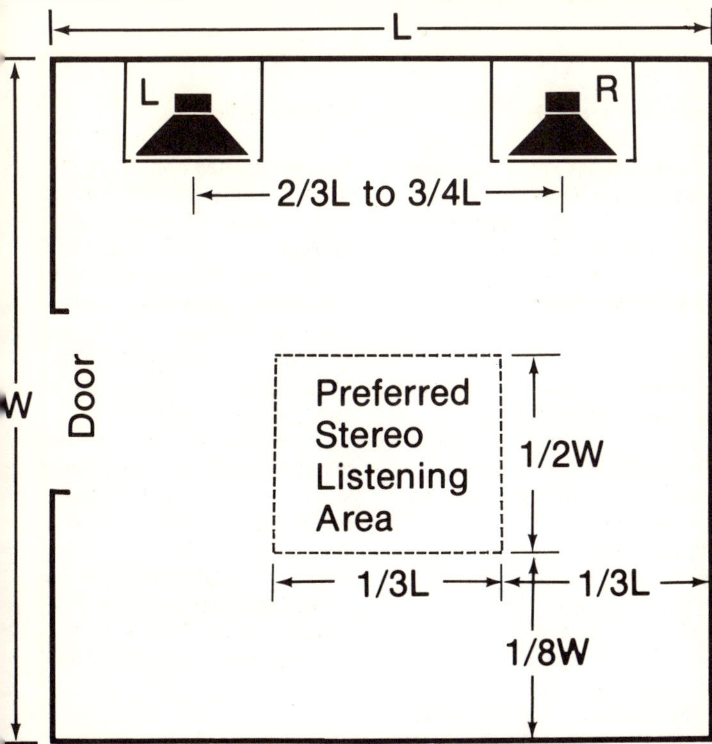

Fig. 10-2 Good stereo sound can be achieved anywhere in the dotted-line area. Dimensions are given in terms of length and width of your room.

next to one of the speakers. If you are a reasonable distance from both speakers (not necessarily the same distance from each) you'll get all the stereo effect you need. Balance controls on amplifiers compensate for differences in perceived level from each stereo channel.

If you've got a rectangular room in which the long dimension is considerably greater than the short one, use one of the long walls for your speakers. Don't place them far apart in the corners. If you do you're liable to get an exaggerated stereo effect which is not musical and is often referred to as the *hole-in-the-middle effect.*

Some manufacturers of amplifiers even supply a *center channel* output intended for connection of a third amplifier (mono) and speaker. The aim is to fill that hole-in-the-middle and preserve the continuous wall of sound effect which stereo is supposed to provide.

Typically, a rectangular room might have the stereo speakers arranged as shown in Fig. 10-2.

Dimensions are given as fractions of wall lengths rather than in absolute terms so you can adapt the diagram to your own room size. (Note that *our* rooms allow for at least one entrance.)

The L-Shaped Room

Whoever came up with the L-shaped living room/dining room combination in contemporary American homes surely did audiophiles no favor. Not only am I saddled with that particular kind of arrangement but one of my walls has a brick fireplace, the opposite wall has an archway leading to the center hall, and the third wall is almost all windows. This arrangement left me with very little choice. I settled for the scheme in Fig. 10-3 and some of the elements in my final arrangement may suggest ideas if you have a similar problem.

My stereo speakers had to go on the *short* wall as shown. They are mounted on the mantle shelf running the length of that wall above the brick fireplace. Distance between the speakers is roughly twelve feet. My favorite listening position is from the long sofa located at the back of the room near the rear archway leading to the hall. This sofa can comfortably seat four people. The large arm-chair shown near the entry to the dining area is too close to the right-channel speaker and the other pair of chairs near the window wall are too close to the left-channel speaker. It couldn't be helped.

As for acoustic treatment: the floor is carpeted and the entire window wall is draped with lightweight fabric. All other walls and the ceiling are "hard" or untreated in any way. The arrangement allowed me easily to switch to four-channel sound a few years ago. My rear speakers are now mounted on either side of the hall archway behind any listeners seated on the sofa. Without the two large archways in the room this minimal treatment would probably have resulted in too "live" or reverberant a listening area. Since sound goes right out the openings without being reflected back into the listening area, the openings make additional treatment unnecessary.

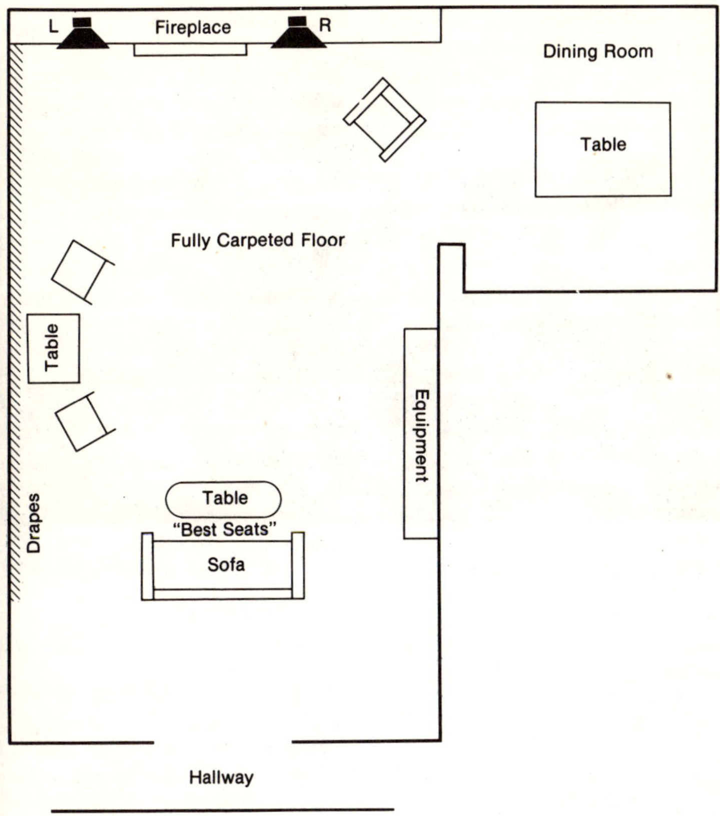

Fig. 10-3 One way to install a stereo system in a problem area.

Solving a tough speaker placement problem.

Tips on Setting Up Your Listening Room

Some general rules applicable to almost any listening room:

1. If you are going to carpet the floor, don't treat the ceiling with acoustic tile or other absorptive material.

2. Only one of each pair of walls should be treated with absorptive material such as draperies.

3. Try not to place your phonograph equipment too close to either speaker. Otherwise speaker vibrations might get back into the tone arm, causing acoustic feedback.

4. Try to provide more than one chair located for good stereo listening. If you don't you'll end up giving the "stereo seat" to your guests and miss hearing your own system properly.

5. If at all possible arrange the equipment (particularly the controls) within arm's reach of your favorite listening position.

6. If the sound you end up with doesn't seem to be as satisfying as that which you heard in the dealer's showroom, don't be afraid to use the tone controls on your receiver or amplifier. Remember: used in moderation they can add just the tonal emphasis you feel is lacking.

Extra amplifier and decoder below were added for a first conversion to quadraphonic sound.

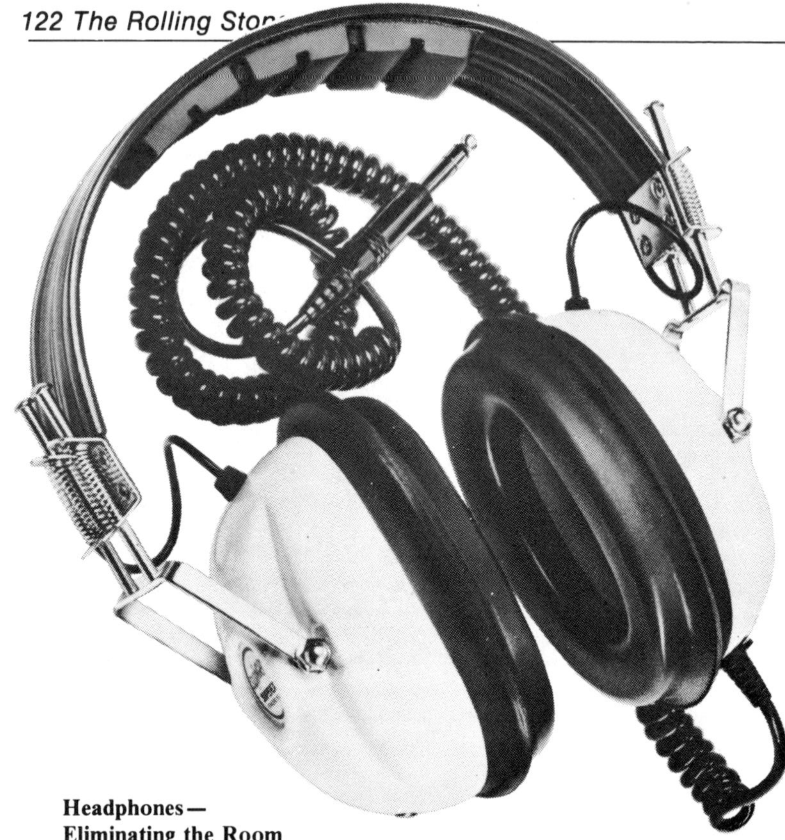

Stereophones, like this Superex Model SW-2, can eliminate the problems of room acoustics.

Headphones— Eliminating the Room

There is one way to listen to stereo programs without considering the influence of room acoustics—through stereo headphones that bring sound to your ears without any intermediate medium. Some people would rather listen to their systems via headphones than loudspeakers. If you've never listened to stereophones, at least give them a try. The feeling you get is not the same as when listening to speakers in an open room. The music seems to be inside and above your head and many people feel that the stereo separation is exaggerated. (Most recording engineers "mix" their product with the loudspeaker-oriented listener in mind.) Nevertheless there are a great number of headphone sets sold each year by at least a dozen companies.

If you do like what you hear while wearing headphones, don't buy the first pair you try. The big thing about headphones (besides the usual considerations of frequency response and power handling capability) is comfort. Although most headphones are adjustable to the head of the wearer, after you have them on for several hours some are going to seem more comfortable than others.

Besides eliminating the acoustics of the listening room, headphones also resist other noises and distractions about you. If you are a first-time listener to phones, chances are you'll talk at shouting level while wearing them. Avoiding this tendency comes only with practice. A final word about phones: if while listening you find that the orchestra is reversed (left to right), there's nothing wrong with the phones or your amplifier. Just take the phones off and turn them around. Many people have missed that simple point!

Test Records

There are some excellent test records around that can help you judge your system very quickly. CBS Records puts out one (Catalog No. STR-101) that has bands for checking overall frequency response of the system, tracking ability of your cartridge and a lot of other important parameters. Other test records offered by some of the hi-fi periodicals from time to time provide samples of "difficult-to-reproduce" musical selections that nicely put your equipment through its paces and quickly show up any deficiencies. I strongly recommend that you get one or more of these records. Play them every so often to make sure that everything is still working as it should. They'll also impress friends who haven't yet been turned on to hi-fi.

Hi-Fi Clinics

There are some companies around who run portable "hi-fi clinics." You bring your equipment to these people and they run some tests using a lot of impressive-looking expensive equipment, give you some printed curves and more often than not try to convince you that your equipment is not performing up-to-snuff and that you'd better trade it in for their "superior" brand. If your system is giving you distortion-free musical reproduction, why bother with this kind of "reassurance"? If something really goes wrong with your system you'll know it before anyone else —no matter how much laboratory equipment they have. Suppose that in fact your "50 watt per channel" amplifier is only putting out 45 watts. If you can set your music loud enough and still not hear distortion, are you really going to worry about "specmanship" at that point?

There's one thing that no stereo system can do for your listening room. Because it is much smaller than a concert hall it can never duplicate the latter experience. If that fact leaves stereo a couple of steps away from "ultimate fidelity," then four-channel or quadraphonic sound attempts to close that final gap.

Chapter 11

FOUR-CHANNEL SOUND

If you think of monophonic sound as one-dimensional listening—listening to music through a peep hole—and if you think of stereo sound as two-dimensional listening, then four-channel sound is full three-dimensional listening. When listening to stereo, you can tell if an instrument is to the left, to the right, or even centered between the two loudspeakers. You cannot, however, get a real feeling of the hall in which a performance was made. Assuming a recording is made in a concert hall, reverberant sound is of course picked up from the microphones. Reproduced by means of a stereo system the original sounds of the instruments plus the reverberance or ambience of the hall are all heard from the front speakers. It was the desire to reproduce the feeling of the hall more realistically that prompted early experiments in four-channel sound.

Four-Channel Sound on Tape

The easiest way to accomplish four-channel sound is on tape. You will remember from Chapter 8 that it is no trick at all to put as many discrete channels as you want on tape. In professional recording studios sixteen-track and twenty-four-track recording is commonly used. To mix down from sixteen tracks to four tracks is no more difficult than mixing down from sixteen tracks to two for stereo, which recording studios had been doing all along. Earliest consumer versions of four-channel recording were in the form of open-reel discrete tapes. It was Vanguard that must be credited with the first actual four-channel tapes. Tape deck producers such as Teac and Sony were quick to get into the discrete open-reel tape format and offer record and playback machines in late 1969 or early 1970. People hearing these tapes for the first time insisted that quadraphonic sound was as much a step forward from stereo sound as stereo was from mono.

A little later on RCA entered the quadraphonic field by introducing hardware and software for eight-track cartridge tapes. First they called the system Quad-8 but since then the name has been changed to *Q-8* cartridges. Instead of having four two-channel programs on an eight-track cartridge, these new Q-8 cartridges offer two quadraphonic programs contained in the eight available tracks. We have already discussed the problems that exist for four-channel cassettes (see "Quadraphonic cassettes," Chapter 8). For the moment, therefore, quadraphonic sound on tape is limited to open-reel (Figs. 11-1 and 11-2) and Q-8 cartridge formats.

Fig. 11-1 A four-channel tape unit intended for home use.

Matrix Phono Discs

In late 1969 Peter Scheiber, a young musician, started getting press coverage when he announced that he had developed a way in which four channels could be compressed, combined or matrixed down to two complex channels. Once the four channels had been compressed to two-channel form you could do everything to them that you could do to any stereo signal. Specifically you could record them on a disc because we have two-channel discs. You could transmit them over the air as stereo FM because we have an approved system for stereo broadcasting. Scheiber also demonstrated a black box that would be used at home to extract four channels from the two compressed channels.

Fig. 11-2 A semi-professional quadraphonic recorder. Young Asylum Record stars, the Eagles (here shown are Randy Meisner, Glenn Frey, Don Henley and Bernie Lendon) use TEAC's 3340 four-channel simul-sync tape deck for rehearsals.

Although he did not disclose his technique at first, other people were working on the same problem: to find a way to make good four-channel records. In this country John Fixler and I worked up such a system which we sold to Electro-Voice Company. This system became known as *Stereo-4* and Electro-Voice began to market adapters which could be connected to an existing stereo system to convert it to four-channel use. Of course any *real* four-channel system requires four separate channels of amplification (the equivalent of two stereo amplifiers). Obviously, it takes four loudspeakers to get four-channel sound as well.

Meanwhile in Japan several companies were also at work on the problem. For a while it seemed that there would be many different systems of four-channel disc recording, each incompatible with the others. Instead two basic matrix systems evolved: the

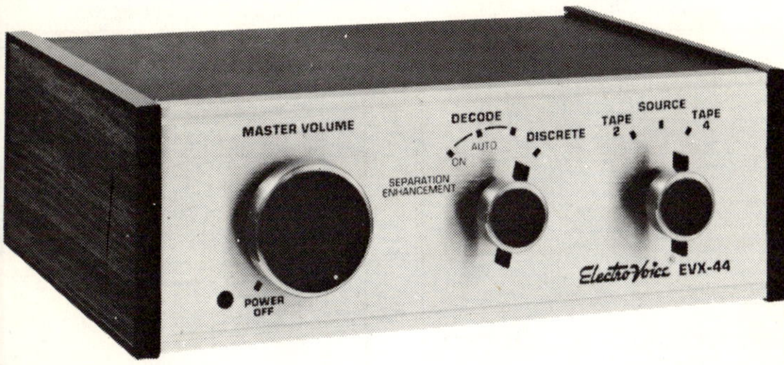

A second-generation matrix disc decoder by Electro-Voice, Inc.

regular matrix, RM, and the *SQ* matrix (developed some time later by CBS). The regular matrix is the simplest to understand and is very close to what Peter Scheiber first proposed:

Starting with four channels of music on four tape tracks, these four channels are mixed together in the following way:

Lt = .92Lf + .38Rf + .92Lb - .38Rb
Rt = .92Rf +.38Lf + .92Rb - .38Lb

The Lt signal is the new left-total signal which will be used in cutting the left groove-wall in the conventional stereo record. The Rt signal is the new right-total signal which will be used in cutting the right groove-wall. Notice that each of these total signals contains varying amounts of all four original signals. The abbreviations for the signals are Lf = left front, Rf = right front, Lb = left back and Rb = right back. Throughout this chapter I will use "b" (for back) instead of "r" (for rear) so as not to confuse the terms "right" and "rear." The minus sign in front of the Rb in the first signal complex and in front of the Lb in the total right signal simply means that those signals have been phase-inverted. Referring to Fig. 11-3(a), you see a sine wave or continuous tone signal. In Fig. 11-3(b) the sine wave has been phase-inverted; that is, it goes in a negative direction

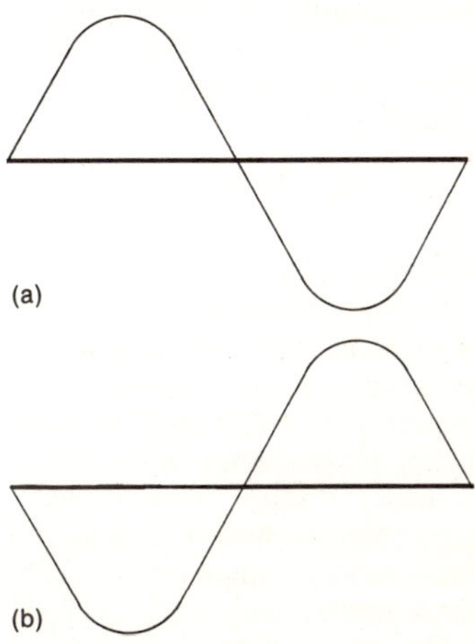

Fig. 11-3 Sine-wave in (b) has been *phase-inverted* with respect to sine-wave signal shown in (a).

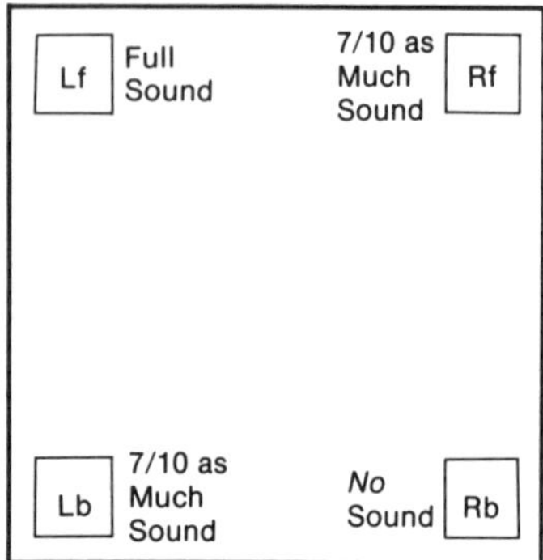

Fig. 11-4 How a *left-only* sound is heard over the four speakers using the Scheiber matrix encode-decode system.

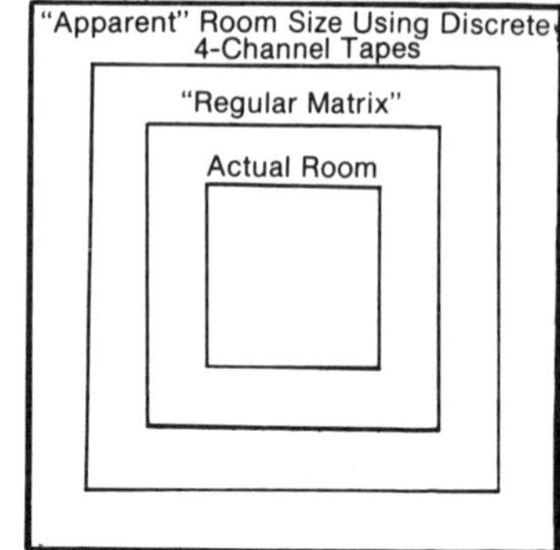

Fig. 11-5 Apparent room size seems to expand symmetrically when listening to regular matrix discs, and even more so listening to *discrete* four-channel sources.

when the first signal is going positive, and in a positive direction when the first signal is going negative. Reversing the phase of a signal is easily accomplished electronically. The various numbers in front of each signal contribution simply indicate the amplitude of each of these signals as compared to its original amplitude of one.

Matrix Decoding

In order to recover four independent signals from this mixture of two signals used to make the recording, Scheiber used the following further mixes in the decoding process:

$L^{1}f = .92Lt + .38Rt$
$L^{1}b = .92Lt - .38Rt$
$R^{1}f = .92Rt + .38Lt$
$R^{1}b = .92Lt - .38Rt$

If we were to expand the first of these equations in terms of the original signals we would come up with:

$$L^{1}f = Lf + .7Rf + .7Lb$$

Notice that the recovered left front signal contains a predominance of the original left front signal plus smaller amounts of right front and left back. Interestingly the original right back signal has been cancelled entirely and is not present in the new left front output signals. If we were to solve for the other recovered signals in terms of the original content, we would get:

$R^{1}f = Rf + .7Lf + .7Rb$
$L^{1}b = Lb + .7Lf - .7Rb$
$R^{1}b = Rb + .7Rf - .7Lb$

In each case the recovered signal contains a predominance of the desired signal and lesser amounts of adjacent channel signals; the diagonally-opposite signal is completely absent. In order to understand what this sort

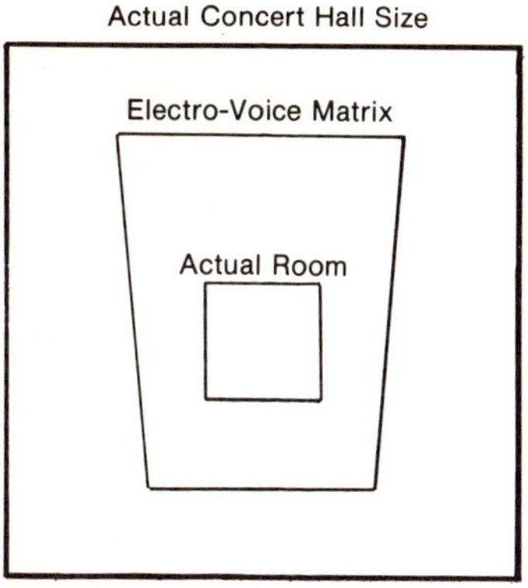

Actual Concert Hall Size

Electro-Voice Matrix

Actual Room

Fig. 11-6 The early Electro-Voice matrix system favored front-back separation at the expense of left-right separation.

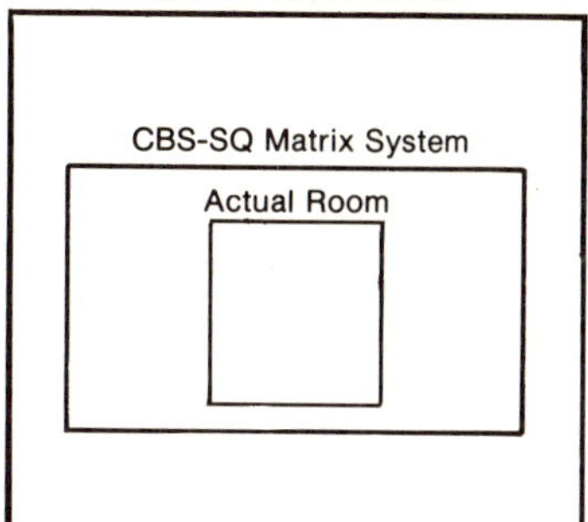

Fig. 11-7 The widely-accepted SQ matrix system emphasizes left-right separation at the expense of front-back separation.

of reproduced sound will be like in a listening room, consider the diagram of Fig. 11-4. In that diagram a left front only signal was encoded and decoded by means of the equations and circuits of the Scheiber system. The left front speaker will reproduce a full-amplitude signal. The left back and right front speakers will also produce that same left front signal but only seven-tenths as loud as it will be heard from the left front speaker. The diagonally-opposite right back speaker will not produce any sound.

A listener sitting at or near the center of the room will hear the sound coming from the left front speaker just where it belongs. If, however, he moves very close to the left back speaker he will think that the sound is coming from that undesired location. Thus the separation capability of the Scheiber matrix approach is limited to about three dB. A decibel, you will remember, is a relative measurement of sound or signal levels. By way of comparison, separation figures achieved with two-channel stereo phonograph pickups are 20 to 25 dB, while stereo FM broadcasts require that 30 dB of separation be maintained between left and right channels.

If we were to consider that a concert hall has the dimensions shown in Fig. 11-5 and if, in that diagram, the next inner rectangle represented our apparent impression when listening to fully discrete four-channel tapes, then the room effect we would get when listening to a system such as the regular matrix would be that of the next inner rectangle. Separation is symmetrical around the four channels. That is, separation is just as good between any two channels as between any other two channels around the room.

The system first used by Electro-Voice is somewhat similar, but the difference in the encoding and decoding equations afford more front-to-back separation. Using a similar analogy Fig. 11-6 shows the apparent room size to the listener with the Electro-Voice system. Notice that front separation from left to right is better than that of the regular matrix system and front-to-back separation is also much better. But back left-to-right separation is poorer than that of the regular matrix.

Finally Fig. 11-7 illustrates the room impression that is gained from the later-developed CBS-SQ encoding and decoding system. Here left-to-right separation is very good both at the front of the listening room and at the rear. Front-to-back separation, though, is minimal. Without getting into the mathematics of the other systems it should be evident that there are any number of ways of encoding four channels of sound into two channels and decoding back to four. The difference between the various systems lies in this choice of where you want maximum separation.

At first it was thought that all these systems would be incompatible. Actually one kind of record played through another kind of decoder gives a very acceptable four-channel effect. The only incompatibility that results is one of instrument placement; that is, if you use one kind of decoder for another kind of matrix disc the various instruments may not sound as though they are coming from precisely where the record producer intended. Since you don't know what the record producer had in mind this is really not too serious a form of incompatibility

The CBS-SQ system and the Sansui QS system use one more trick in encoding and decoding. The extra ingredient is known as 90-degree phase-shifting and it is intended to further increase apparent audible separation. As we saw in Fig. 11-7 the CBS system stresses left-to-right separation—separation is as good from left to right as in any stereo record. Regular matrix and the somewhat similar Sansui QS systems stress diagonal separation.

Logic Systems

It is possible to get a good four-channel effect from simple matrix decoders available today. More elaborate decoders or decoder circuits use something called *logic* or *gain-riding* to enhance the limited separation inherent in all matrix systems. Putting it simply, a logic or gain-riding system detects which channel is supposed to be the loudest at any given instant. Acting on this information the amplifying channels change their amplification instantaneously. Thus the channel that should be playing loudest plays louder while the channels

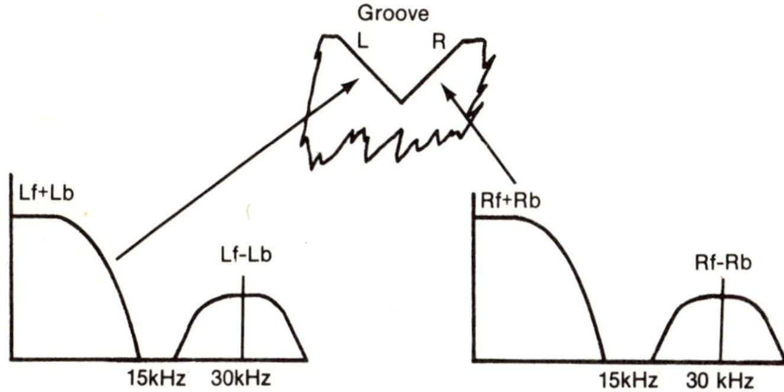

Fig. 11-8 A single groove in a Quadradisc contains the four complex audio signals plotted in this diagram.

that should be subdued become even more subdued. There are many ways to accomplish this logic or gain-riding function and each manufacturer has developed his own circuitry for this idea. Sansui calls its logic system *Variomatrix*. Makers of sophisticated decoders for the CBS-SQ system call it *double-logic* or *full-logic*. The gain-riding works for front-to-back separation and a second circuit works to improve side-to-side separation and diagonal separation. From the purist's point of view no matrix system—however good the associated logic circuitry—can ever produce quite as much separation as can a discrete system of tape or disc.

Discrete Four-Channel Discs

While the various matrix disc proponents were fighting it out for dominance, JVC (Japan Victor Company) developed a method for actually cutting four different signals in a single record groove. They called this new kind of disc *CD-4* and in June 1971 offered the first CD-4 records for sale along with associated playback equipment. During their first year of existence the sale of these discs and equipment was primarily limited to the Japanese home market. RCA rejected all the various matrix techniques and insisted that the only route to four-channel discs was the discrete one. Early in 1972, RCA announced that working with its affiliate JVC it had perfected the CD-4 disc principle sufficiently to warrant full commitment to this system.

The make-up of this new discrete disc groove can best be understood by looking at Fig. 11-8. In that diagram each wall of the groove is considered to contain its own composite signal. The left wall contains an audio signal extending from 30 Hz to 15 kHz, but that audio consists of the *sum* of left-front and left-back programs. In addition a high-frequency carrier having its center frequency at 30 kHz is modulated by the difference between Lf and Lb (Lf-Lb). This difference signal actually varies the high-frequency carrier above and below its center frequency. The highest frequency of the sub-carrier can reach 45 kHz while the lowest frequency reached by the sub-carrier is 20 kHz. Modulation of this carrier is therefore FM in nature.

The right wall is similarly recorded except that in this case the regular audio impressed is the sum of right-front and right-back while the audio information used to modulate the high-frequency carrier is right-front minus right-back.

If you were to play such a recording on regular stereo equipment, the left channel would contain left-front plus left-back and both programs would come out of your left-front speaker. The right-front speaker would reproduce right-front plus right-back. The record is therefore compatible in stereo because all four channels are heard even though they are mixed together as far as front and back sounds are concerned. In order to reproduce this kind of record in full four-channel sound, it is necessary to have something called a *demodulator*.

The first job of the demodulator is to recover those difference signals from the high-frequency carriers. This job is similar to that performed by an FM tuner which detects the audio from the constantly changing frequency of the radio frequency carrier. Once the two difference signals (one from each groove wall) are recovered they can be remixed with the sum signals that have been recorded using conventional audio frequencies. Mixtures are done in accordance with the following equations:

$$(Lf + Lb) + (Lf - Lb) = 2Lf$$
$$(Lf + Lb) - (Lf - Lb) = 2Lb$$
$$(Rf + Rb) + (Rf - Rb) = 2Rf$$
$$(Rf + Rb) - (Rf - Rb) = 2Rb$$

Notice that the left-front signal has been recovered with no mixture of other signals and the same is true for the other three desired signals. This record has been called a discrete four-channel one because the original four signals are recovered individually with no mixture between them.

Because this kind of record contains frequencies extending all the way up to 45,000 Hz an ordinary stereo cartridge is not likely to be able to reproduce these records; most such cartridges perform well up to about 20,000 Hz. Manufacturers of phono cartridges began to develop new designs that would have good frequency response all the way out to 45,000 Hz. Many of these are available already as are the demodulators necessary for recovering the four independent signals. Even the design of the stylus tip required some rethinking. If you were to visualize how closely spaced the wiggles are when you consider 45,000 back-and-forth-wiggles-per-second in a record groove you will realize how difficult the job of the tracking stylus must be if it is to follow these rapid alternations. A new stylus shape was developed which is neither elliptical nor conical. It is called a Shibata stylus after its Japanese inventor. This stylus tip tends to hug more of the groove wall surface and can therefore trace these rapid 45,000 Hz alternations more accurately. The nice thing about CD-4 records is that even if you are not ready to buy four-channel equipment you can play them on good stereo equipment without damaging the four-channel program content. If at some later date you decide to convert your system to four channels these CD-4 records will reproduce music in full four-channel sound. RCA calls these records Quadradiscs in this country.

Discrete Four-Channel Broadcasting

As we have already pointed out, all matrix-encoded discs can be broadcast over stereo FM stations. No special equipment is required since the discs are really a form of two-channel recording. In the case of CD-4 or Quadradisc records, the problem is much greater. These discs cannot be directly broadcast nor can they first be decoded and then broadcast as four separate audio signals.

It would take new rules and regulations set by the Federal Communications Commission to permit quadraphonic discrete broadcasting. A committee composed of industry members called the NQRC (National Quadraphonic Radio Committee) has been formed to investigate several proposals for broadcasting four separate signals over a single FM station. The first system proposed was developed by Louis Dorren, a young West Coast engineer. Since he first made his proposal, actual on-the-air tests have been performed with the permission of the FCC and produced very good four-channel results.

Unfortunately this system (as well as most of the other systems currently being considered by the NQRC and the FCC) requires a greater channel bandwidth than is required by either mono or stereo FM broadcasting. Among the many considerations that will have to be evaluated by the FCC is that of possible interference from one station to another. In many of the proposed systems the SCA service discussed in Chapter 5 would have to be shifted in frequency. This would mean changing or replacing the thousands of special receivers now in the hands of subscribers to background music services.

The FCC is not likely to jump to any hasty conclusions. Extensive field tests are scheduled for early 1974 and it is likely to be some time beyond that before the FCC makes a decision—if in fact it ever does. It took over three years for the FCC to evaluate the various systems proposed in the late 1950s for stereo broadcasting. The only thing manufacturers can do right now (and many of them are already doing it) is to equip the back panels of their quadraphonic receivers or their FM tuners with a special detector output. This extra jack might one day be used for connection to a quadraphonic FM adapter. The new complex signal yet to be approved would be fed to this adapter and out would come four independent signals. In this sense, such an adapter of the future might be compared to the CD-4 demodulators we discussed earlier. No manufacturer is likely to build such an adapter at the present time since he has no way of knowing which of the ten or more systems currently being considered is likely to be the winner. Besides Louis Dorren's Quadracast system, such giants as RCA, Zenith and General Electric are all competing with variations on his innovation.

Converting to Four-Channel Sound

Let's assume you now own a decent stereo system and you want to upgrade it to include the latest advances in four-channel sound. First, of course, you will need a second pair of speakers. The question always raised is: "Must the rear channel speakers be the same as the original front stereo pair?"

Some dealers will tell you that it is perfectly okay to have lower quality speakers in the rear channels. If all you wanted to listen to in four-channel was the ambience effect of the concert hall this advice would probably be valid. But if you have heard any of the new quadraphonic records (either matrixed or discrete), you know that composers and arrangers are taking fullest advantage of the new medium, placing instruments on all four channels and creating primary musical effects from each. This is especially so in popular or rock music where instruments literally surround you and you become part of the performance. Under such circumstances all speakers should be of equal quality. For that matter the extra pair of amplifiers (usually a second stereo amplifier) should have as much power capability as the amplifier in your original stereo set-up. If your original speakers are no longer manufactured, the next best thing to do is to select a pair of speakers whose sound most nearly approximates that of the original pair.

In addition to the extra speakers and extra double amplifier you will need a decoder. If you're interested in CD-4 records you will need a demodulator as well. To date no one has offered a combined decoder-demodulator as one accessory package. A typical hook-up of a converted stereo unit is shown in Fig. 11-9. The decoder uses the tape monitor interruption jacks as a means of getting in and out of the old receiver or amplifier. While this uses up the tape-monitoring facil-

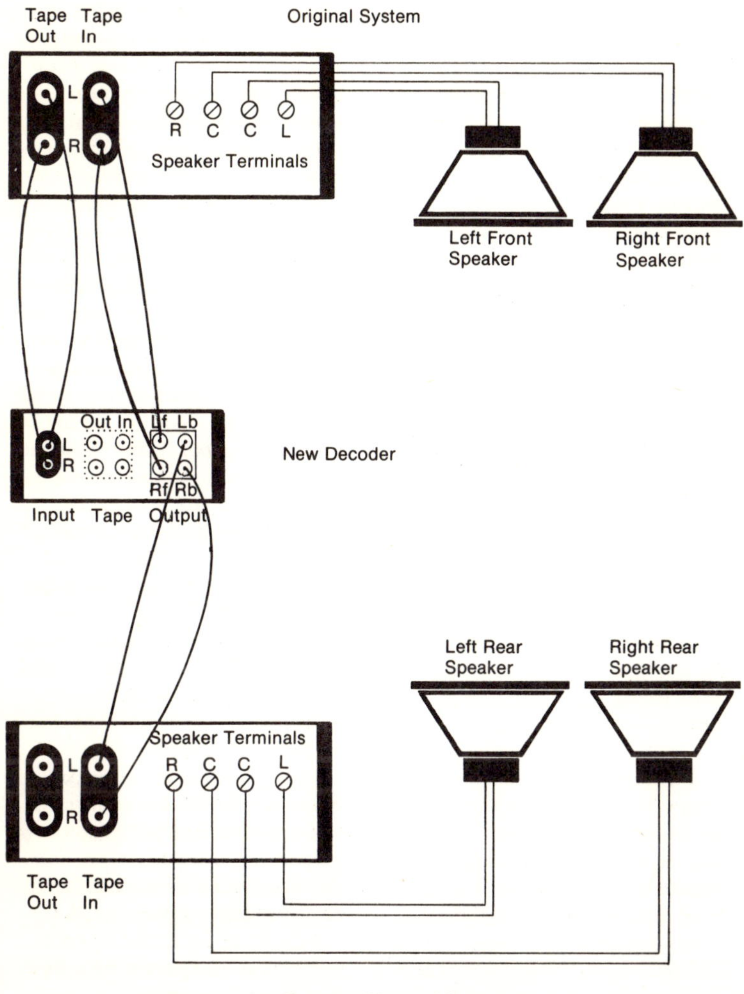

Fig. 11-9 How to connect a matrix decoder between an existing stereo system and the extra amplifier and speakers required for quadraphonic sound.

ities—assuming there is only one set of tape monitor input and output jacks—on your primary piece of equipment, most decoders repeat the tape monitor tape facilities on *their* back panels. This enables you to have tape-monitoring again.

Figure 11-9 shows the set-up when using a matrix decoder. Fig. 11-10 shows a slightly different arrangement for incorporating a CD-4 demodulator. Note that in the case of the discrete CD-4 set-up the phono cartridge is connected directly to the demodulator and the outputs of the demodulator are connected to the auxiliary inputs of the old receiver and of the new extra amplifier. Thus a CD-4 demodulator connection need not use up tape-monitoring facilities of your old equipment. As a matter of fact such a conversion to CD-4 leaves you with a bit of unused equipment: the preamplifier section—the one that amplifies the very low level signals from your phono pickup—is unused in your old receiver. That function is now accomplished by the new CD-4 demodulator. If you are considering CD-4 or Quadradiscs you will need a new phono cartridge specifically designed to be able to reproduce these kinds of records. It will also play your older stereo records very well—perhaps better than your stereo cartridge did.

Your stereo record collection has not been made obsolete by the advent of four-channel sound. Most stereo discs played through a matrix decoder set-up such as that shown in Fig. 11-9 will actually produce a sort of four-channel effect. You will hear sounds coming from the back speakers that are different from the sounds coming from the front speakers. This is because there are many out-of-phase components in the two audio signals originally recorded on stereo discs. According to the encoding and decoding equations shown earlier these out-of-phase components go through the decoder and get added up in such a way that they become back-channel signals. The degree of four-channel effect that you get will vary from stereo record to stereo record. Virtually every old stereo disc will give you *some* effect and most people prefer it to stereo listening.

In connecting your new speakers to the extra amplifier it is important that they be connected in phase with each other and that these new speakers are also operating in phase with the front stereo speakers. It is easy enough to properly connect front speakers in phase when both are connected to the same amplifier. When two amplifiers are used it is possible that the speakers connected to the first amplifier may very well be out-of-phase compared to the two speakers connected to the separate second amplifier.

Merely following the positive and negative symbols on the amplifier and speaker terminals may not insure equal phasing of

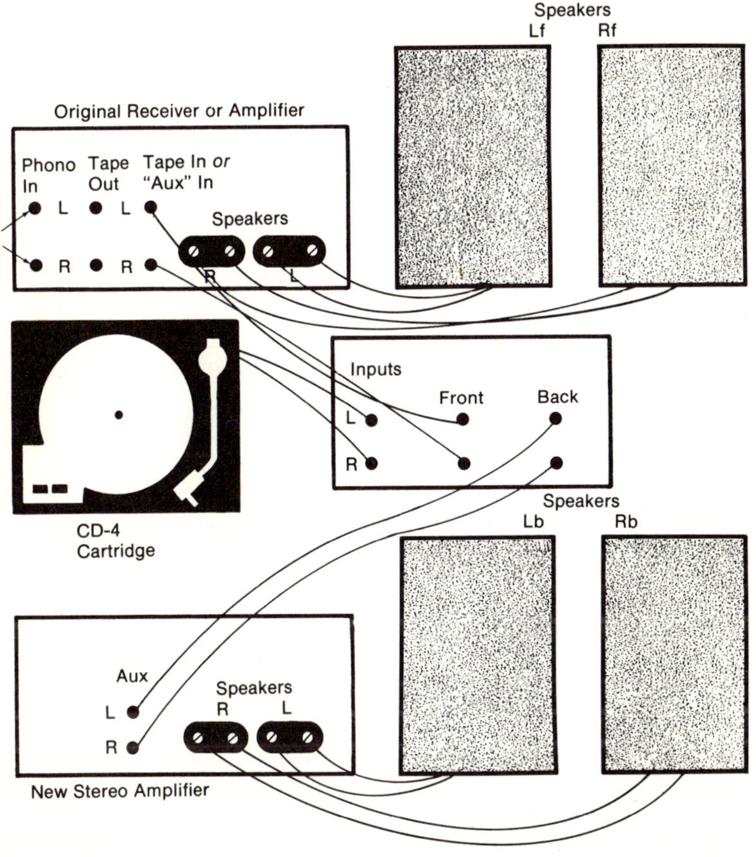

Fig. 11-10 Connecting a CD-4 demodulator in converting a stereo system to quadraphonic.

front and back speakers. The easiest way to check for phasing between front and rear is to turn down the volume control (in some cases the balance controls may have to be used) of, say, both left channels. Face the right side of the room as if the right-front and right-back speakers constituted a stereo pair. Then throw your entire system into mono and listen for good centered bass-response between them. If the centered image seems vague or if the bass seems deficient, reverse the connection to the right-back speaker. You can then repeat the entire experiment on the left side of the room by turning down the volume controls to both right speakers and concentrating on the left-front and left-back channels as if they constituted a stereo pair.

While the diagrams of Figs. 11-9 and 11-10 indicate that you will end up with quite a few components, it is not as bad as it seems. Many of the new decoders have a master volume control and some even have tone controls for the rear channels. The decoder becomes a sort of control center much like the preamp control chassis discussed earlier. Some manufacturers offer combination stereo-amplifiers-plus-decoders. These are intended to perform all the electronic functions you need in converting to four-channel sound. The use of such a combination amplifier-decoder reduces the number of separate components required.

New Four-Channel Equipment

If you are lucky enough to be starting from scratch, there is a great deal of equipment available for putting together a really fine quadraphonic sound system. The very first efforts by manufacturers to produce all-in-one quadraphonic receivers usually contained four separate channels of amplification, a conventional FM stereo section and a simple matrix decoder suitable for SQ or RM decoding. In general the power output capability of these receivers was lower than that of the previous popular stereo receivers in the same price class. This is simple to understand since a great deal more circuitry has to be included in the chassis. Four 10-watt-per-channel amplifiers actually contain more circuitry than two 20-watt-per-channel amplifiers. In addition some form of decoder circuit has to be included. So the cost of quadraphonic receivers seemed discouraging.

Even so, four 10-watt amplifiers equal 40 watts. When listening to quadraphonic sound with all four channels playing at once, you do get the sense of having the equivalent of 40 watts of amplifier power; perhaps even more so than if it consisted of two channels each going at 20 watts maximum.

Nevertheless the popular low-efficiency speaker systems pose a problem with these low-powered receivers. Some speaker manufacturers are therefore developing

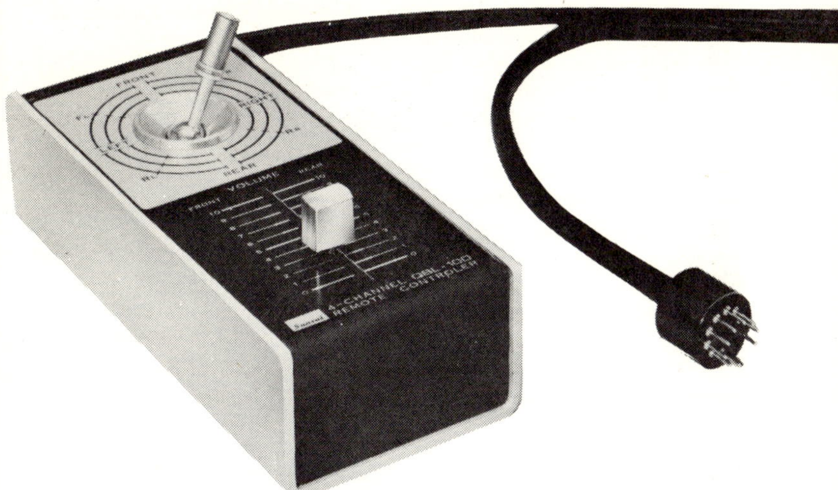

The remote-control attachment for Sansui's four-channel receivers lets you balance and adjust sound from your listening chair.

more efficient designs that would require less input power to produce good sound levels.

Hedging just a little bit many manufacturers came up with another idea for quadraphonic receivers. Realizing that prospective customers might want to buy a receiver now with only two speakers but would want four-channel capability for the future, they developed circuits called *bridged amplifiers* or *strapped amplifiers*. In receivers of this type it is possible to throw a switch that effectively parallels pairs of amplifiers to produce more than twice the power-per-channel in stereo use compared to the power in quadraphonic use.

For example, a receiver which might be capable of producing 10 watts on each of its four channels could be expected to produce as much as 25 watts per output channel when operated in the strapped stereo mode. Such a receiver can be purchased now with only two speakers and used for stereo playback for as long as you like. Whenever you wish, you can buy a second pair of speakers, connect them, flip the two channel/four channel switch and away you go in full quadraphonic sound.

Starting in 1973 many manufacturers began to offer both built-in CD-4 and built-in matrix decoding in single-piece complete receivers. Few of the 1973 receivers offered full-logic or even partial-logic enhancement of the

A quadraphonic stereo receiver with built-in CD-4, SQ and RM sources, by Pioneer.

matrix decoding. The cost of including CD-4 plus matrix decoding plus logic plus reasonably high power was just too prohibitive. In 1973 integrated circuits (ICs) were just not developed to a point where they could be used to help reduce the cost involved. Building all this circuitry using individual components would make the unit excessively large and excessively expensive. Some receivers offered in 1973 contained many choices of matrix decoding plus logic; they generally did not include CD-4 circuitry. Other receivers incorporated CD-4 demodulators but only simple matrix decoding without logic circuitry. If you would be perfectly satisfied with matrix decoding minus any logic or gain-riding circuitry you need wait no longer. If you change your mind later there will surely be separate decoders on the market that will be as sophisticated as you might want and will contain even better logic circuitry for matrix decoding. You can also start out with a receiver that does not even contain CD-4 demodulator circuitry and add that as a separate accessory later on.

The Future of Four-Channel Sound

Many people at first looked upon four-channel sound as a new gimmick dreamed up by manufacturers to sell more speakers and more electronic equipment. By now most listeners have to agree that four-channel sound, properly handled, adds a new dimension to home music listening. It appears that matrix and CD-4 discs will coexist for a long time to come. Receivers are likely to include more and better decoding circuitry for matrix discs. They will also probably incorporate improved demodulator circuitry for discrete discs. Both of these kinds of circuits will ulti-

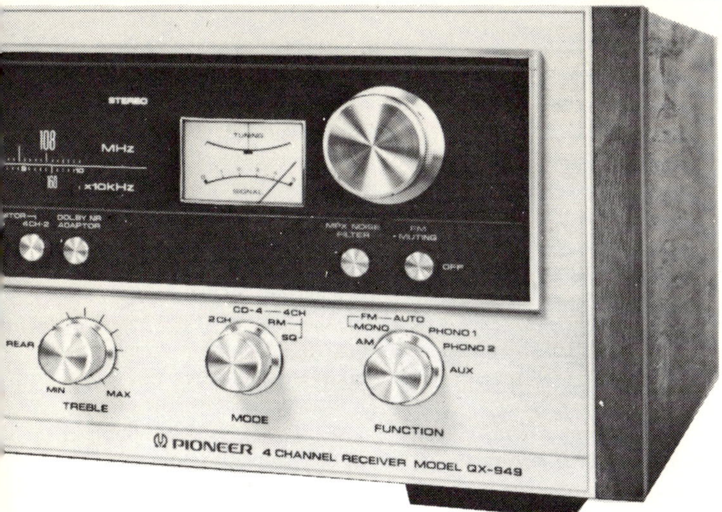

mately be reduced to ICs—and prices are likely to come down soon. Once that happens manufacturers will probably try to increase the power capability of each amplifier channel in such receivers. No doubt the FCC will act some day to give us four-channel discrete FM broadcasting as well.

The Future of High Fidelity

Progress in high fidelity home music reproduction has been enormous during the last twenty-five years. The massive technical-looking pieces of equipment of the post World War II era gave way to more attractively styled mono equipment in the '50s, compact good-looking stereo equipment in the '60s and four-channel equipment in the '70s. What next? Eight-channel octaphonic? Sixteen channels?

Probably quadraphonic sound will become and remain the dominant system of home music reproduction for the next decade. Record makers are quickly discovering that by using one of the four-channel techniques it is possible to place sounds anywhere in three-dimensional space. Of course more refined placement of such sounds could always be achieved by the addition of more channels—more sources of sound. Such additional channels would not improve the sonic perspective nearly as much as any of the giant leaps from mono to stereo or quadraphonic was able to do.

Most likely the next few years will see improvements in circuitry, miniaturization of circuits and possible breakthroughs in the area of loudspeaker design. Meanwhile the home music system of today is truly superior in performance to the professional sound systems of just a few years ago.

Glossary

A-B Test: A method of evaluating the performance of similar high fidelity components, especially speaker systems, by switching rapidly from one to the other.

AC: Alternating current. Often used as synonym for electrical house current.

Acoustic Feedback: Unwanted low-frequency acoustic interaction between output and input of an audio system, usually between loudspeaker and microphone or turntable cartridge.

Acoustics: The science of sound. Also used to refer to the acoustical character of halls and rooms.

AF (Audio Frequency) Frequency within the range of human hearing—approximately 20-20,000 Hz.

AFC (Automatic Frequency Control): An AFC circuit of an FM tuner corrects for some inaccuracy in tuning by locking in the station being tuned.

Air Suspension (Acoustic Suspension) Speaker Systems: Speaker system design which permits good bass reproduction from relatively small-sized enclosure in which sealed volume of enclosed air aids in determining speaker suspension characteristics. Generally low efficiency design, requiring higher amplifier power than "vented" or ported designs.

AM (Amplitude Modulation): Modulation accomplished by varying the amplitude or intensity of a high frequency signal called the carrier frequency.

AM Suppression: Ability of an FM tuner to suppress changes of amplitude in received signals, thereby improving the signal-to-noise ratio by rejecting noise and interference. Also helpful in reducing "multipath" effects.

Ambience: In quadraphonics, a reference to reverberant sound as opposed to sound coming directly from musical instruments.

Ampere: Basic unit of current.

Amplification: Increase in signal magnitude.

Amplifier: Component providing amplification of signal, from low to higher voltage or current (pre-amplifier) or power (power, basic or main amplifier).

Antenna: Assembly of metal bars, wires or loops for picking up radio waves. Dimensions depend largely on the type of signal to be received.

Anti-Skating Device: Mechanism exerting a small outward force on a tone arm to compensate for the inward thrust caused by tone arm geometry.

Automatic Turntable (Record Changer): Record player which can change records automatically.

Aux (Auxiliary) Input: Input on amplifier, etc. Accepts extra signal source such as cassette tape player, TV sound, etc.

B

Baffle: The board on which one or several loudspeakers are mounted. Separates the sounds radiated from the front and back of the speaker.

Balance Control: Control used to adjust volume difference of left and right stereo channels or front to back volume difference in four-channel equipment.

Bass: Low audio frequency range, below approximately 200 Hz.

Bass Reflex: Loudspeaker enclosure with an opening permitting sound from the rear of the speaker cone to be radiated from the front of the enclosure.

Beat: A pulsation caused by interaction of two waves of different frequencies.

Bias: An electrical signal of relatively high frequency applied to magnetic tape during the recording process, along with the audio signal. The bias frequency is generally in the range of 70 to 120 kHz.

Binaural: Two-eared. Sometimes used erroneously to mean stereophonic.

Block Diagram: A schematic diagram illustrating the main circuit blocks and signal flow in an electronic device.

C

Capacitor: A device which can store an electric charge or block DC voltages from getting to a circuit.

Capstan: Tape drive spindle in a tape deck.

Capture Ratio: An FM tuner's ability to reject unwanted FM stations and interference occurring on the same frequency as the desired station.

Carrier: Main radio signal from a transmitter. Can be modulated in different ways (AM, FM) to convey sound information at the receiving end. Also high frequency signal used in CD-4 "discrete" four-channel records.

Cartridge: Phono pickup or endless loop tape in a packaged, standardized container.

Cassette: Preloaded container with tape and spools for use on cassette tape recorder. A miniature reel-to-reel tape system.

CD-4: A phonograph record that can store four channels of discrete sound. See *Quadradisc.*

Center Channel: An output terminal found on some stereo amplifiers which supplies a monophonic mixed L + R signal.

Channel: A complete sound path. Monophonic or mono systems are single channel. Stereophonic or stereo systems use two channels usually identified as L (left) and R (right). Quadraphonic sound uses four channels.

Channel Separation: Degree to which left and right channel signals are separated in a stereo pickup: FM stereo tuner, amplifier, etc.

Coaxial Cable: A cable consisting of an inner conductor and an outer shield. Used as antenna input leads and for interconnecting audio units.

Compatibility: Multiple use of program source equipment. Examples include: FM MPX signal receivable as mono by radio or tuner; stereo record properly played with a mono pickup; stereo pickup properly playing mono records or four-channel record capable of being played in stereo or mono.

Compliance: The flexibility of stylus motion in a phono cartridge. The unit of compliance is 10^{-6} cm/dyne.

Component System: An audio system consisting of separate units.

Conical Stylus: A stylus whose tip is circular.

Continuous Power (RMS Power): Most conservative method of specifying power output in an audio amplifier. Involves use of a single, continuous test tone fed to amplifier while output power is measured to predetermined distortion levels.

Crossover Frequency: In loudspeaker systems the borderline frequencies between low/medium range and medium/high range speakers. The point at which two bands of frequencies are separated, such as the frequency at which mid-range and tweeter receive the same amount of power from the amplifier.

Crosstalk: Amount of right channel signal heard in left channel, and vice versa. Expressed as level of unwanted signal in relation to wanted signal channel, measured in dB.

Cutting Stylus: Stylus used for cutting of phonograph records.

D

Damping Factor: Ratio of loudspeaker impedance to amplifier's internal impedance. Describes amplifier's ability to damp unwanted residual speaker movement.

dB (decibel): A logarithmic unit used to express the ratio between two acoustical or electrical power, voltage or current levels. Mathematically:

$$dB = 20 \log_{10} \frac{\text{voltage}_2}{\text{voltage}_1} \text{ or}$$

$$dB = 10 \log_{10} \frac{\text{power}_2}{\text{power}_1}$$

A unit of the relative intensity of sound or amplitudes of two electrical signals. Used for comparing the intensities of two different signals (e.g., one may be 10 dB greater than the other), or expressed with reference to some fixed arbitrary level. A difference of one dB is generally considered the smallest that can be differentiated by the human ear. A 3 dB difference in level is clearly heard by almost everyone.

Decoder: In an FM stereo tuner, the circuit that extracts the left and right channel signals from an FM MPX broadcast signal. Also a device that extracts four-channel sound from two-channel encoded sound.

De-Emphasis: Attenuation of high sound frequencies in an FM tuner, to counteract the boosting of these frequencies ("pre-emphasis") done at the FM station's studio.

Demodulator: Circuit used to extract four channels from a CD-4 discrete disc.

Derived Center Channel: See *Center Channel.*

Derived Four-Channel Sound: A method of obtaining a kind of four-channel sound from two-channel sound sources, such as stereo records.

DIN (Deutsche Industrie Normen): German Industrial Standards. In audio, the German standard of plugs, sockets, etc.

Direct-Radiating Speaker Systems: Speaker designs in which all speaker elements (drivers) are mounted on front surface and radiate sound directly at the listener, as opposed to omnidirectional designs in which one or more speakers may be directed towards reflecting surfaces.

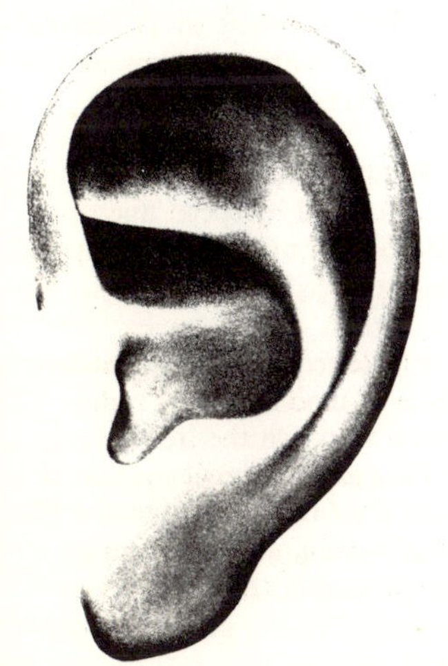

Discrete: Four-channel sound; quadraphonic sound handled as such without conversion to two-channel; four independent sound sources on tape or disc played back via two stereo amplifiers into four speakers.

Dispersion: Distribution of sound from a speaker over an angle into a listening room.

Distortion: Output signals not present in the original input. Any departure from the original. See *Harmonic Distortion, IM Distortion.*

"Dolbyized" Tape: A prerecorded tape which was made using the Dolby noise reduction process. Such tapes should be played back only on machines equipped with Dolby playback circuitry if accurate frequency response is desired. Played back on ordinary tape machines, Dolbyized tapes will tend to sound shrill and exhibit an over-abundance of high frequencies.

Drift: Tendency of a tuner to move away from optimum adjustment as its components become warm.

Dubbing: Copying of previously recorded material. In tape recording playing a recorded tape on one machine while recording it on another.

E

Efficiency: In loudspeakers, the percentage of electrical input converted to acoustical output.

Electrostatic Speaker: A loudspeaker utilizing the principle of a two-plate capacitor in an electrical field.

Elliptical Stylus: A pickup stylus shaped so that its width across the groove is greater than the width of its sides.

Enclosure: A cabinet which houses one or more speaker units.

Encoder: A matrix circuit for combining four sound channels into two.

Equalization: Correction for frequency non-linearity of recordings. Phonograph records are cut with low frequencies attenuated and high frequencies boosted. Playback equalization compensates for this, producing a flat frequency characteristic.

F

FET (Field Effect Transistor): Special kind of transistor consisting of metal oxides. Used in FM and other audio equipment because of its good linearity and high input impedance.

Filter: A circuit which attenuates signals above, below or at a particular frequency.

Flutter: Quick changes of pitch caused by speed fluctuations in the movement of tapes or discs.

FM (Frequency Modulation): Type of modulation of radio waves in which the frequency, not the amplitude, of the carrier is altered by the audio signal. FM broadcasting achieves higher sound quality than AM.

Four-Channel Stereo: An audio technique using four (instead of two) channels for sound reproduction.

Frequency: The number of complete cycles of a wave per second. The basic unit is the Hertz (Hz). Multiples are the kiloHertz (thousand Hertz) and the megaHertz (million Hertz). KiloHertz is abbreviated as kHz; megaHertz as MHz.

Frequency Response: The frequency range which a unit will reproduce or to which it should respond.

Front End: Section of a tuner that selects the desired signal from the radio band and converts the RF signal to IF.

Function Selector: In an amplifier or receiver, the switch or knob which selects the different program sources such as phono, FM, AM, etc.

G

Gain: Degree of signal amplification achieved in an amplifier circuit.

H

Harmonic Distortion: The sum of all signals in an output which are multiples of the input signal frequencies ("harmonics"). Expressed as a percentage of the total output.

Head Alignment: Adjustment so they are at right angles to the longitudinal axis of the tape. Also known as azimuth alignment.

Headphones: Small transducers, usually mounted in pairs on a bracket to fit over the head and designed to make intimate contact with the ears. Also known as earphones or stereophones.

Hertz (Hz): Unit of frequency equal to one cycle per second.

Hole-in-the-middle Effect: Discontinuity in the desired stereophonic "wall of sound" or "stage." Generally caused by having stereo speaker systems spaced too far apart or by playing program source with exaggerated stereo separation.

Horn: Speaker unit using a flaring funnel to join its sound vibrations to the surrounding air.

Hum: Unwanted low-frequency tone. Usually caused by power line frequency of 60 Hz and its harmonics.

Hysteresis Motor: A motor used in high quality turntables. Rotates at constant speed regardless of power voltage fluctuations.

I

IC (Integrated Circuit): Solid circuit "chip" containing the functions of numerous transistors, diodes, resistors, etc.

IF (Intermediate Frequency): The frequency which results in a tuner when the incoming signal from the antenna is mixed with the local oscillator frequency.

IHF (Institute of High Fidelity): Institute founded by American manufacturers of audio equipment, devoted to the improvement of audio technology, standardization of test and measuring methods, etc. "IHF" in audio specifications means that values were obtained in measurements made according to IHF standards.

Image Rejection: The ability of a tuner to reject an RF signal which appears at an incorrect point on the radio dial.

IM Distortion: Intermodulation distortion. (See *Distortion.*) Signals in output caused by interaction of two or more input signals but not harmonically related to them. Expressed as a percentage of total signal output.

Impedance: Resistance to the flow of alternating current. Measured in ohms.

Induced Magnet: A phono cartridge in which both magnet and coils are fixed. The moving part is a tiny piece of iron.

Induction Motor: Type of drive motor used on less expensive turntables and record changers. Subject to speed variations when power line voltage changes. Generally lower in cost than synchronous types.

Infinite Baffle: Type of loudspeaker mounting involving a totally sealed enclosure.

Instantaneous Peak Power (IPP): A meaningless power rating developed by advertisers to make their products "look better" in terms of power output. Can be as much as twenty times as great as the continuous power, depending upon who is doing the lying.

Integrated Amplifier: Unit combining a preamplifier and power amplifier.

IPS: Inches per second. A term used to describe the movement of tape.

L

Limiter: Circuits in an FM tuner that reject unwanted amplitude variations caused by atmospheric or vehicular ignition noise, or other man-made electrical interference signals.

Logic Circuit: Circuits used to augment the otherwise minimal separation inherent in "matrix" encoded four-channel discs. Most are based upon variable-gain amplifiers (gain-riding) which detect which signal is instantaneously dominant and adjust the decoder's parameters dynamically to favor that channel.

Loudness Control (Contour): A circuit which counteracts our reduced sensitivity to very low and high notes at lower than "live" volume levels.

Low Filter: A filter circuit designed to remove low-frequency noises from a program.

M

Magnetic Cartridge: A phono cartridge which derives its electrical output signal from changes effected in a magnetic circuit.

Main Amplifier (Power Amplifier): Amplifier unit which produces the output power required for driving speakers.

Masking Effect: Psycho-acoustical phenomenon in which low level sounds are obscured or "masked" by the presence of loud sounds. Principle is used in a variety of audio applications, notably in the Dolby noise reduction process.

Matrix: A circuit used for the addition and subtraction of signals. The circuit used for encoding four related sound sources into two channels on tape or disc, requiring a matrix decoder to retrieve the original four channels.

Microvolt: Millionth of a volt.

Monaural: One-eared. Sometimes used erroneously to mean monophonic.

Monitoring: Listening to sounds, during the recording process, to judge or control the sound quality.

Monophonic: Recording, transmission and reproduction of sound using a single channel.

Moving Coil Cartridge: Magnetic phono cartridge in which the coils move and the magnet is fixed.

Moving Magnet Cartridge: Magnetic phono cartridge in which the magnet moves and the coils are fixed.

Multipath Reception: Arrival of FM signal via several paths of different lengths, due to reflecting objects.

Multiplex: Transmission of two or more channels on a signal carrier so they can be independently recovered by the receiver. In FM stereo: transmission of L + R (sum) signal and L - R (difference) signal on main carrier and subcarrier, respectively. Abbreviated as MPX.

Multiplex Decoder: See *Decoder.*

Music Power: The maximum power available over a short period of time from a power amplifier. Also called "dynamic-power" and IHF Music Power.

N

Noise: Unwanted signal consisting of a mixture of random electrical signals. Also the sum of all unwanted signals such as hum, hiss, rumble, interference, distortion, etc.

O

Ohm: Basic unit of resistance or impedance.

Omni-Directional: Equal sensitivity or output in all directions. Used to describe antennas, microphones and loudspeakers.

Oscillator: An electronic circuit which generates an alternating current.

Output Impedance: Impedance at output terminals of a device such as an amplifier or tuner, etc.

Output Stage: Final stage of a power amplifier which supplies power to a loudspeaker.

Overloading: Feeding into a system a signal that exceeds the system's capability.

P

Phase Lock Loop: A circuit which can be used for accurate frequency stability in FM tuner circuits or stereo multiplex decoders. In operation, feedback voltages are used to "lock" a local frequency generator to a reference signal, such as the incoming radio signal, etc. More accurate than Automatic Frequency Control (AFC) used in the past for similar applications.

Phasing: Connections between power amplifier and speakers in a stereo system made in such a way that signals representing a central sound source cause the speakers to move together, i.e., in phase.

Power Amplifier: See *Main Amplifier.*

Power Bandwidth: The frequency range over which a power amplifier will produce half of its rated output power (according to IHF standard).

Power Handling Ability: Maximum amount of power than can be safely fed into a loudspeaker.

Preamplifier: A circuit unit which takes a small signal from a tuner or turntable, and amplifies it sufficiently to be fed into the power amplifier for further amplification.

Pre-Main Amplifier: An integrated audio amplifier consisting of a preamp and a power amp.

Q

Quadradisc: RCA's name for CD-4 discrete records.

Quadraphonic: Four-channel stereo.

Q-8: RCA's name for four-channel, eight-track tape cartridges.

QS: A matrixing technique for encoding four-channel sound into two channels; developed by Sansui Company.

R

Rated Power Output: The maximum power that an amplifier will deliver without exceeding its specified distortion rating.

Ratio Detector: Circuit in an FM tuner for extracting audio signals from modulated radio or intermediate frequency signals.

Receiver: See *Tuner-Amplifier*.

Regular Matrix (RM): A four-channel disc recording and playback system developed in Japan in which four channels are encoded down to two for recording or broadcast purposes and decoded back to four when played through suitable home decoder. Symmetrical in its separation capability from any one channel to the others. QS matrix system, developed by Sansui Company, is a variation of Regular Matrix.

Resistor: Circuit device which offers resistance to the flow of electric current. Resistors may be made from wire, metallic film, carbon and other materials.

Resonance: The tendancy of mechanical or electrical devices to resonate at a particular frequency.

Response: See *Frequency Response.*

Reverberation: Reflection of sound from walls or ceilings.

RF (Radio Frequency): The frequency of a radio carrier wave. AM occupies frequencies between 535 and 1605 kHz, FM occupies those between 88 and 108 MHz.

RIAA (Record Industry Association of America): Usually refers to the disc recording and replay frequency response curves established as standards by this assocation.

Rumble: Low-frequency noise resulting from vibrations in platter and motor of a turntable.

Selectivity: The ability of a tuner to receive only the desired station while rejecting stations which are close in frequency. Measured in decibels (dB).

Sensitivity: The input signal level required by a tuner, amplifier, etc., to be able to produce a stated output. The lower the necessary input, the better the sensitivity.

Separation: See *Channel Separation.*

Shibata Stylus: Stylus developed to play discrete four-channel records.

Signal-To-Noise Ratio (S/N ratio): Ratio of desired signal voltage to unwanted noise and hum voltage. A high S/N ratio is desirable. Expressed in decibels (dB).

Solid State: Circuits using semiconductors, such as transistors and integrated circuits (ICs).

Speaker System: Enclosure containing one or more speakers and a crossover network.

SQ: A matrixing technique for encoding four-channel sound into two-channel. Developed by CBS Records.

T

Tape Deck: Tape equipment comprising complete tape transport system as well as pre-amplifiers for recording and playback, but no power amplifier or speakers.

Tape Hiss: High frequency random noise characteristic of recording tape, often intensified by the electronics used during the record and playback process in tape recording.

Stereophonic: Recording, transmission and reproduction of sound via two independent channels.

Stylus: A finely-machined piece of sapphire or diamond. The part of a phono pickup that traces the record groove.

Synchronous Motor: Type of AC electric motor in which motor speed is related directly to frequency of power source.

Tape Monitoring: Listening to tape-recorded results a fraction of a second after recording is made. Possible only with three-headed tape player/recorders, in which separate playback head "monitors" recorded results by feeding signal to circuit-interruption point in amplifier or receiver. Tape monitor switch on amplifier or receiver permits use of this facility as well as the connection of other audio accessories such as decoders, graphic equalizers, reverberation units, etc.

Tape Print-Through: The magnetization of a section of tape from a layer of tape immediately above or below it. Can be caused by excessively strong recorded signals. The symptom of print-through is a weaker sound following the original sound almost immediately.

THD (Total Harmonic Distortion): See *Distortion.*

Tone Control: Control circuits used to vary the proportion of bass and treble in the sound.

Trackability: Ability of phono cartridge to track record grooves of high amplitude and velocity. Also see *Compliance.*

Tracking Error: Deviation of center-line of phono cartridge from tangency to record groove at point of stylus contact. Caused by tone arm geometry.

Tracking Force: Downward force of playback stylus when playing a record, generally specified by turntable and cartridge manufacturers in grams. In general, lower tracking forces result in less record wear over many playings. Often confused with "tracking pressure," which depends upon tracking force as well as area of surface contact.

Transducer: Device for converting energy from one form to another.

Transient: Sudden change in signal amplitude as caused by percussion instruments, plucked strings, etc.

Transient Response: The ability of an amplifier, cartridge or speaker to follow sudden changes in the level of a sound.

Tuner: The part of a receiver (or a separate unit) which receives radio broadcasts and converts them into audio frequency signals.

Tuner-Amplifier: Unit combining the functions of a tuner, preamplifier and power amplifier. Also called "Receiver."

Tweeter: A speaker designed to reproduce the high frequency portion of the sound spectrum.

V

Vent: Opening or port in a bass reflex loudspeaker enclosure.

Voice Coil: A coil or wire attached to a speaker cone, through which electrical audio signal is passed, causing magnetic action and cone movement.

Volt: Basic unit of electrical pressure or electromotive force.

W

Woofer: A speaker designed to reproduce the low part of the sound spectrum, e.g., organ, bass, etc.

Wow: Slow variation of pitch caused by speed fluctuation in tape or record movement.

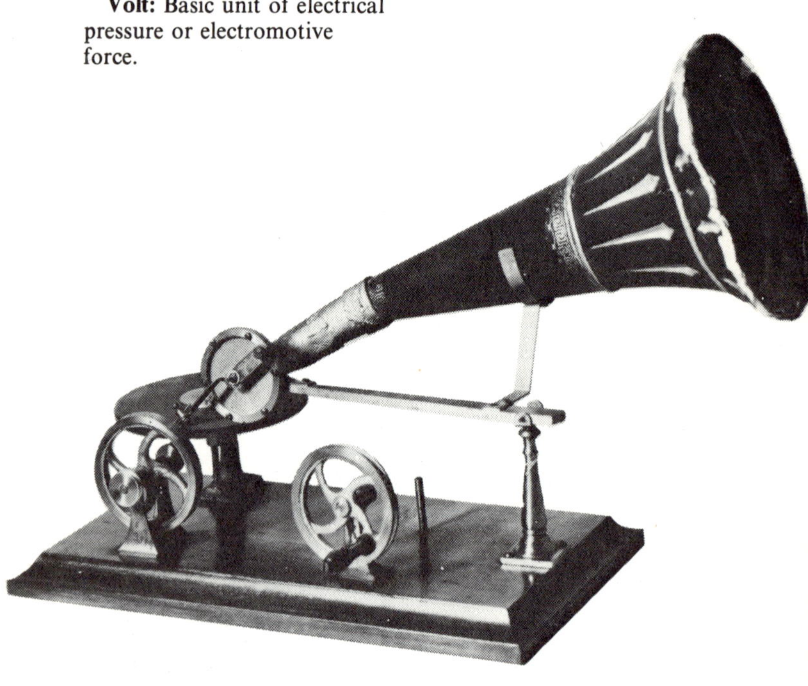

Book design by Brent Beck

Illustration Credits
The Bettmann Archive: 1, 6, 7, 10, 142, 143, 145 (top), 148, 151, 153, 154, 155, 157, 158, 160; Culver Pictures: 8, 11, 144, 145 (bottom), 146, 149, 150, 152, 156, 159

Straight Arrow Books: Victoria Jackson, Carol Raskin-Ward, Bond Francisco, Roger Carpenter, Larry Miller, Bill Cruz, Wendy Werris, Linda Gunnarson, Rosemary Nightingale, Jon Goodchild and Alan Rinzler